Assessing the Implications of Policy Options for the Military Personnel Budget

An Analytic Framework for Evaluating Costs and Trade-Offs

MATTHEW WALSH, THOMAS LIGHT, RAYMOND E. CONLEY

Prepared for the Department of the Air Force
Approved for public release; distribution unlimited

 PROJECT AIR FORCE

For more information on this publication, visit **www.rand.org/t/RRA1218-1**.

About RAND

The RAND Corporation is a research organization that develops solutions to public policy challenges to help make communities throughout the world safer and more secure, healthier and more prosperous. RAND is nonprofit, nonpartisan, and committed to the public interest. To learn more about RAND, visit www.rand.org.

Research Integrity

Our mission to help improve policy and decisionmaking through research and analysis is enabled through our core values of quality and objectivity and our unwavering commitment to the highest level of integrity and ethical behavior. To help ensure our research and analysis are rigorous, objective, and nonpartisan, we subject our research publications to a robust and exacting quality-assurance process; avoid both the appearance and reality of financial and other conflicts of interest through staff training, project screening, and a policy of mandatory disclosure; and pursue transparency in our research engagements through our commitment to the open publication of our research findings and recommendations, disclosure of the source of funding of published research, and policies to ensure intellectual independence. For more information, visit www.rand.org/about/research-integrity.

RAND's publications do not necessarily reflect the opinions of its research clients and sponsors.

Published by the RAND Corporation, Santa Monica, Calif.
© 2023 RAND Corporation
RAND® is a registered trademark.

Library of Congress Control Number: 2022918756
ISBN: 978-1-9774-1014-6

Cover: Carol Ponce (composite design); adapted from image by Benjamin Faske/U.S. Air Force.

About This Report

The military personnel budget provides financial resources to compensate active military personnel (MILPERS). This includes pay and allowances, subsistence of enlisted personnel, permanent-change-of-station travel, and other military personnel costs.

Spending on MILPERS has grown at an average annual rate of 3.3 percent per year since fiscal year (FY) 2000, to approximately $36 billion in FY 2021. Growth in MILPERS spending at this rate threatens to undermine readiness and crowd out future efforts to modernize key military capabilities.

To help better manage and control spending on MILPERS, the deputy chief of staff for Manpower, Personnel and Services (AF/A1) needs tools to understand the future cost implications of workforce and personnel policies that the Air Force might pursue. This report describes and applies one such tool that can be used to evaluate the implications of workforce and personnel policies on MILPERS spending. The tool integrates models that simulate how policy changes might reshape the personnel inventory with a derivative pay table that contains elements that make up standard composite pay rates. The tool can be used to translate workforce and personnel policies into changes in the personnel inventory and changes in MILPERS spending. A series of simulation studies show that certain personnel policies would allow the U.S. Air Force to save tens of millions to hundreds of millions of dollars annually without making changes to pay and allowances, end strength, or grade strength. Larger levels of savings require additional changes beyond personnel policies.

The research reported here was commissioned by the director of Manpower, Organization and Resources, Headquarters U.S. Air Force (AF/A1M) and conducted within the Workforce, Development, and Health Program of RAND Project AIR FORCE as part of an FY 2021 project "Manpower/Personnel Realignment Tool."

RAND Project AIR FORCE

RAND Project AIR FORCE (PAF), a division of the RAND Corporation, is the Department of the Air Force's (DAF's) federally funded research and development center for studies and analyses, supporting both the United States Air Force and the United States Space Force. PAF provides the DAF with independent analyses of policy alternatives affecting the development, employment, combat readiness, and support of current and future air, space, and cyber forces. Research is conducted in four programs: Strategy and Doctrine; Force Modernization and Employment; Resource Management; and Workforce, Development, and Health. The research reported here was prepared under contract FA7014-16-D-1000.

Additional information about PAF is available on our website:
www.rand.org/paf/

This report documents work originally shared with the DAF on February 7, 2022. The draft report, issued on February 9, 2022, was reviewed by formal peer reviewers and DAF subject-matter experts.

Acknowledgments

We thank our research sponsor, the Office of Manpower, Organization and Resources. We thank Brig Gen Gentry W. Boswell, director of Manpower, Organization and Resources and deputy chief of staff for Manpower, Personnel and Services, Headquarters U.S. Air Force, for his support throughout the project. We also thank Col Patrick White and Col James Barger; this research benefited greatly from their input and support. In addition, we are grateful to Lt Gen Brian Kelly, deputy chief of staff for Manpower, Personnel and Services, for sharing his insights and connecting us with different stakeholders. We thank the Air Force financial management stakeholders whose views informed the research conducted for this project, especially those from the Financial Management and Comptroller.

Finally, we thank the many RAND colleagues who helped with this work—principally, but not exclusively, C. Ben Gibson, Lisa Harrington, Kirsten Keller, Ignacio Lara, Nelson Lim, Miriam Matthews, Libby May, Al Robbert, and Dulani Woods.

Summary

The military personnel budget provides financial resources to compensate active military personnel (MILPERS). This includes pay and allowances, subsistence of enlisted personnel, permanent-change-of-station travel, and other military personnel costs. This report describes and applies a tool that can be used to evaluate the implications of workforce and personnel policies on MILPERS spending.

Issues

- Spending on MILPERS has grown at an average annual rate of 3.3 percent per year since fiscal year (FY) 2000, to approximately $36 billion in FY 2021. This outpaced growth in prices in the overall economy, which averaged 1.9 percent per year for the same period.
- To ensure a ready workforce without undercutting modernization efforts, the DAF must explore options to maximize MILPERS affordability.
- At the same time, the DAF must consider the nonmonetary trade-offs and risks that these options entail.
- The DAF needs an analytic framework to view savings, trade-offs, and risks of different solution options alongside one another.

Approach

To understand how various factors affect MILPERS spending, we reviewed relevant bodies of literature, policy, and DAF documents. We also analyzed financial and end strength data contained in military personnel budget documents to understand how and why MILPERS costs have varied over time and among services.[1] Finally, we developed an analytic framework that integrates existing personnel inventory models, funded authorizations, and personnel cost factors. The framework provides a view of how these factors interact, and it enables simulation of different courses of action (COAs) to alter MILPERS spending while considering affordability objectives and other goals. We demonstrate the analytic framework for multiple simulations chosen based on discussions with the research sponsor that involve changing end strength, grade strength, and experience levels.

Key Findings

- The DAF MILPERS budget is developed by multiplying the estimated work years in various end strength subcategories by discrete cost elements and by summing the totals.

---

[1] End strength is the number of people in a particular grade (and at a particular year of service) at a moment in time, often measured at the beginning or end of a month or FY.

The MILPERS budget can be controlled by reducing end strength or the average cost of an airman.

- Basic pay makes up much of the standard composite pay rates. Solutions that involve limiting the rate of growth of basic pay and other personnel costs might require coordination across the military services and Congressional approval.
- Grade and years of service affect basic pay along with several other elements of the standard composite pay rates. Personnel policies that alter these factors could reduce the cost of an airman, and they would not require Congressional approval.
- The DAF's average cost of an active duty person exceeds those of the U.S. Army and U.S. Marine Corps. This is because of the greater share of officers and the more-senior grade mix in the DAF. Solutions that involve reducing the ratio of officers to enlisted personnel or shifting toward a less-senior grade mix could reduce the average cost of an airman.
- Certain personnel policies would allow the DAF to achieve annual savings of tens of millions to hundreds of millions of dollars annually.
- Larger levels of savings require changes to compensation, end strength, or grade strength.

Recommendations

- **The DAF could use a simulation capability like the one described in this report to link workforce and policy changes to MILPERS spending.** This capability could be used to explore how personnel policies might be used to decrease MILPERS spending. In addition, this capability could be used to explore how personnel policies designed to meet nonfinancial objectives will affect future MILPERS spending.
- **Improve the fidelity and breadth of the simulation capability.** The analytic tool described here establishes a basis for continuing lines of development. These include capturing year-over-year dynamics produced by changes in workforce and personnel policy, increasing the fidelity of monetary outcomes represented in the analytic tool, and incorporating additional nonfinancial outcomes in the analytic tool.
- **Refine and evaluate solution options to reduce MILPERS spending.** Several solution options showed considerable potential to reduce future MILPERS costs. In particular, the DAF should reconsider personnel requirements for platforms, missions, and operations and examine ways to apply workforce and personnel policies in a targeted manner tailored to characteristics of different career fields.
- **Develop solution options with input from operations, plans, programs, financial management, logistics, engineering, and force protection communities.** Approaches for reducing MILPERS spending might introduce risk throughout the DAF enterprise. The problem cannot be solved in the silo of manpower, personnel, and services. Additional perspectives from across the Air Staff and Secretariat are needed to identify risks associated with different solution options and to develop mitigating COAs.

Contents

Figures and Tables

Figures

Tables

Chapter 1. Introduction

Background

The military personnel budget provides financial resources to compensate active military personnel (MILPERS). This includes pay and allowances, subsistence of enlisted personnel, permanent-change-of-station (PCS) travel, and other military personnel costs. The Department of the Air Force (DAF) executes the MILPERS program, and it must balance fielding the workforce structure needed to deliver capability and capacity to today's combatant commanders while ensuring the financial flexibility needed to design and field the future workforce. In addition, the DAF must provide competitive pay and incentives to recruit and retain a talented all-volunteer workforce. There is tension between these goals—reducing compensation might create financial flexibility but make it more challenging to recruit and retain individuals. As the DAF noted in its MILPERS program budget documentation for fiscal year (FY) 2023, "Our biggest leadership challenge is taking care of people while striking the right balance between maintaining today's readiness and posturing future modernization and recapitalization priorities."[2]

In FY 2021, DAF end strength equaled nearly 335,000 active duty officers, cadets, and enlisted personnel.[3] The MILPERS budget provides the pay and allowances for these individuals. The MILPERS budget is large. At a cost of approximately $36 billion, the MILPERS budget made up more than 20 percent of the DAF's total FY 2021 budget of $168 billion.[4] The MILPERS budget is growing. Between FY 2000 and FY 2021, spending on active duty personnel grew at an average annual rate of 3.3 percent per year, outpacing growth in prices in the overall economy, which averaged 1.9 percent per year for the same period.[5] However, despite the size and growth of the MILPERS budget, the active duty workforce shrank by about 20,000

---

[2] DAF, *Fiscal Year (FY) 2023 Military Personnel Appropriation*, Washington, D.C., April 2022.

[3] End strength is the number of people in a particular grade (and at a particular year of service [YOS]) at a moment in time, often measured at the beginning or end of a month or FY. FY 2021 end strength (as of September 30, 2021) equaled 64,936 officers, 266,451 enlisted personnel, and 4,098 cadets (335,485 total) at a cost of $35,862,533,000. Average work years for FY 2021 equaled 349,460; see DAF, 2021.

[4] The FY 2021 active duty MILPERS budget totaled $35,862,533,000 and made up about 21 percent of the DAF's total budget of $168,237,000,000; see DAF, 2021.

[5] The FY 2000 active duty MILPERS budget totaled $17,978,193,000; see DAF, *FY 2002 Amended Budget Submission*, June 2001. Growth in prices in the overall economy between FY 2000 and 2021 is measured using the gross domestic product (GDP) price deflator; see Office of the Under Secretary of Defense (OUSD), *National Defense Budget Estimates for Fiscal Year 2022*, (Green Book), Washington, D.C., August 2021c. A considerable amount of growth in MILPERS spending is concentrated between FY 2000 and FY 2003 and between FY 2016 to FY 2021. Growth from FY 2000 to FY 2003 was in part because of the perceived inadequacy of military pay given the ramp-up in overseas operations that occurred after the terrorist attacks on September 11, 2001.

individuals between FY 2000 and 2021.[6] After normalizing for the size of the workforce, the average cost of an airman increased by 106 percent from $50,000 in FY 2000 to $103,000 in FY 2021. By comparison, prices in the U.S. economy (as measured by GDP price deflator) and civilian pay grew by only 48 percent and 60 percent, respectively, during the same period.[7]

Continued growth in the average cost of an airman in excess of general inflation will create affordability and readiness challenges and could crowd out future efforts to modernize key military capabilities. The chief of staff of the Air Force (CSAF) identified fiscal challenges, including force structure and manpower, in December 2020, saying

> No matter what happens with the budget, it will require us to make tough choices. We need to continue developing a lethal and affordable force that Congress supports. Action Order D drives the Air Force to "make force structure decisions in Fall 2020 and amend force planning processes to create the fiscal flexibility required to design and field the future force we need."[8]

In response to this concern, the CSAF and DAF FY 2023 investment review requested that efforts be made to identify opportunities to reduce workforce costs and better meet current and future challenges.[9] To assist AF/A1 in its designated role as the office of primary responsibility for this effort, RAND Project AIR FORCE (PAF) developed an analytic tool to simulate the cost implications of manpower and personnel policies that the DAF might adopt.

Conceptual Framework for Examining Policies to Limit Growth in the MILPERS Budget

The DAF MILPERS budget is developed by multiplying the estimated work years in various end strength subcategories by discrete cost elements and by summing the totals. MILPERS spending can be controlled by reducing end strength or by reducing the average cost of an airman. For example, by limiting growth in military compensation or by shifting labor to more-affordable workforce segments (e.g., reducing the ratio of officers to enlisted personnel or adopting a more-junior grade mix), the same number of work years could be maintained but at reduced cost. The savings can be used to limit future MILPERS spending, or they can be repurposed. In this section, we present a conceptual framework to explore the solution trade space.

---

[6] FY 2000 end strength equaled 355,654; see DAF, 2001.

[7] Rates of GDP price and civilian pay growth were calculated by RAND from Table 5.1 and Table 5.5 of OUSD, 2021c. Although the civilian pay growth index does include the cost of certain civilian personnel benefits, the composition of those benefits differ from those provided to active duty military personnel, making it not directly comparable.

[8] Charles Q. Brown, *CSAF Action Orders to Accelerate Change Across Air Force*, Arlington, Va.: Chief of Staff, U.S. Air Force, December 2020.

[9] DAF, "Memorandum for MAJCOM/CCs, Field Command/CCs, Input Sources, and Air Force Corporate Structure Members: Addendum 23-014 (Functional Optimization for Affordability)," Washington, D.C., September 2020.

The total cost of the MILPERS budget depends on the size (i.e., average strength) and makeup (e.g., grade strength, experience, and career field mix) of the personnel inventory. The total size of the inventory is driven by demand. Manpower authorizations are determined using systematic processes to specify the human resources needed to conduct DAF missions. A reduction in strength caused by a change in authorizations translates to reduced personnel resources to units and a concomitant reduction in mission capabilities.[10] The composition of the personnel inventory is partially driven by policy (e.g., promotion timing) and endogeneity (e.g., the strength of the economy). The size and makeup of the actual inventory differs from the authorized workforce structure because of constraints in the DAF human capital management system, along with external factors that affect recruiting and retention.

The total cost of the MILPERS budget also depends on the average cost of an airman. This cost partially reflects the makeup of the inventory. The rate of basic pay for officers is higher than for enlisted personnel, and the rate of pay is higher for individuals of more-senior rank. As a result, changes in the composition of the inventory have implications for personnel costs. The cost of an airman also partially reflects personnel policies. For example, basic pay depends on rank and YOS. As a result, policies that shift to a more-junior YOS profile will reduce personnel costs. Finally, the cost of an airman reflects such external factors as annual pay increases approved by Congress.

The conceptual framework shown in Figure 1.1 represents the following dependencies:

- Authorized workforce structure—or the number of individuals authorized to serve by grade and DAF Specialty Code (AFSC)—and personnel policies influence the makeup of the personnel inventory.
- The makeup of the personnel inventory influences the average cost of an airman.
- The size of the personnel inventory and the average cost of an airman can be used to estimate the total cost of the MILPERS budget.
- External factors might further shape the personnel inventory and the cost of an airman.

---

[10] Albert A. Robbert, Lisa M. Harrington, Louis T. Mariano, Susan A. Resetar, David Schulker, John S. Crown, Paul Emslie, Sean Mann, and Gary Massey, *Air Force Manpower Determinants: Options for More-Responsive Processes*, Santa Monica, Calif.: RAND Corporation, RR-4420-AF, 2020.

Figure 1.1. Framework for Evaluating Solutions to Affect MILPERS Spending

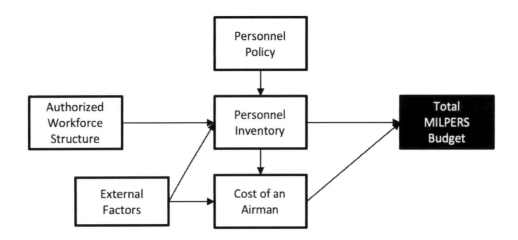

The framework is part of a larger dynamic system. For example, authorized workforce structure depends on inputs from functional areas, the DAF's multilevel corporate structure, and Congress; the cost of an airman depends on annual changes to pay and allowances; and personnel policies have additional costs outside the MILPERS budget (e.g., recruiting and training). Notwithstanding these simplifications, the framework provides a useful guide for examining and comparing the savings and implications of different approaches to limiting the growth of the MILPERS budget.

This report documents an integrated simulation tool built around this framework. The tool links changes in workforce structure and personnel policy to changes in the personnel inventory and, ultimately, to future MILPERS spending. The tool leverages a variety of models developed by RAND PAF to simulate promotion and retention.[11] The tool combines the outputs of these inventory models with a derivative pay table, based on elements of standard composite pay rates, to translate personnel inventories into MILPERS spending.[12]

Preliminary analysis conducted using the tool suggests that by combing multiple personnel policies (e.g., changes in high year of tenure [HYT] requirements and policies affecting retention), the DAF can achieve future savings in the range of tens of millions of dollars to hundreds of millions of dollars annually without changing compensation, end strength, or grade strength. However, larger levels of savings will require changes to compensation, end strength, or grade strength. We explore and summarize some of these options in this report.

---

[11] Matthew Walsh, David Schulker, Nelson Lim, Albert A. Robbert, Raymond E. Conley, John S. Crown, and Christopher E. Maerzluft, *Department of the Air Force Officer Talent Management Reforms: Implications for Career Field Health and Demographic Diversity*, Santa Monica, Calif.: RAND Corporation, RR-A556-1, 2021.

[12] Department of the Air Force Instruction (AFI) 65-503, *Financial Management: US Air Force Cost and Planning Factors*, Washington, D.C., July 13, 2018.

Research Approach

To develop a framework for evaluating solutions to limit future MILPERS spending, we relied on a multimethod approach that included: (1) an extensive review of the existing literature on MILPERS costs; (2) quantitative analysis of historical DAF MILPERS cost and manpower data and a comparison of DAF MILPERS spending in FY 2021 with MILPERS spending in the other military services; (3) quantitative analysis of accession, promotion, and retention trends among DAF servicemembers; and (4) computational simulations using officer and enlisted inventory projection models. Below, we summarize each of these approaches.

Review of MILPERS Costs

AFI 65-503 contains operating and support cost definitions and planning factors for the DAF. The RAND Corporation research team collected tables from AFI 65-503 that report separate elements of standard pay rates and other personnel costs. In addition, the team reviewed guidance from the Office of the Secretary of Defense pertaining to military pay and allowances.

Quantitative Analysis of MILPERS Costs

The research team gathered MILPERS funding and end strength data contained in military personnel budget documents from FY 2000 to FY 2021, as well as historical data on the separate elements of the standard pay rates reported in AFI 65-503. The team conducted analyses to understand factors that historically contributed to growth in DAF MILPERS spending. The team also collected and analyzed data on MILPERS spending by each of the other military services in FY 2021.

Quantitative Analysis of Accession, Promotion, and Retention Trends

The research team gathered officer and enlisted personnel data from the Military Personnel Data System. The team used these data to estimate promotion timing, promotion rates, separation rates, and other inputs needed for the inventory projection models.

Computational Simulations

The research team leveraged an officer inventory projection model to determine how workforce and personnel policies could shape future personnel inventories. In addition, the team created a new model to simulate the effects of workforce and personnel policies on the enlisted inventory. The team used these models, along with a derivative pay table that incorporates elements of standard composite pay rates, to translate workforce and personnel policies to changes in MILPERS spending. The policies were chosen based on discussions with the research.

Organization of the Report

The remainder of this report is organized as follows:

- Chapter 2 describes the components of the active duty standard composite pay rates and how these are combined with a measure of work years to determine spending on MILPERS.
- Chapter 3 describes how DAF MILPERS spending has changed over time and how it compares with MILPERS spending by the other services in FY 2021.
- Chapter 4 describes an analytic tool for simulating how workforce and personnel policies change spending on MILPERS.
- Chapter 5 demonstrates the analytic tool using eight case studies that involve wide-ranging workforce and personnel policies.
- Chapter 6 summarizes findings and lists recommendations.

Chapter 2. Drivers of MILPERS Spending

DAF MILPERS spending can be estimated using the size and grade mix of the workforce and standard composite rates reflecting the cost of an airman. AFI 65-503 Table A19-2, "Active Air Force Standard Composite Rates by Grade," provides the annual military pay rates used to estimate the cost of military personnel in planning and other efforts. In this chapter, we describe the components of standard composite pay rates published by the Deputy Assistant Secretary for Budget (SAF/FMB) each year. We also discuss how the total size of the workforce, as measured by work years distributed across officer and enlisted grades, is derived for the purposes of our analysis. Finally, we link personnel policies to separate elements of standard composite rates that they might influence.

Standard Composite Rates for Active Duty Personnel

The standard composite pay rates published by SAF/FMB as part of AFI 65-503 are the prescribed cost factors to be used for "cost studies, economic analyses, component cost analyses, military construction projects, Program Objective Memorandum inputs, as well as programming, budgeting, accounting, and recording payments from other government agencies."[13] They are calculated according to the provisions of Volume 11A, Chapter 6, Appendix I of the *Department of Defense Financial Management Regulation*.[14] We augment the standard composite pay rates published by SAF/FMB to allow for variation in basic pay and the retirement pay accrual (RPA) by YOS in addition to grade, as described below.

Elements of Standard Composite Pay Rates

Standard composite pay rates include the following elements: basic pay, retired pay accrual (a percentage of basic pay), basic allowance for housing (BAH), Medicare-eligible health care accrual, basic allowance for subsistence (BAS), special and incentive pay, PCS, and miscellaneous pay.[15] Collectively, basic pay, BAS, and BAH accounted for 67 and 64 percent of standard composite pay rates for officers and enlisted personnel, respectively, in FY 2021.

---

[13] DAF, 2018.

[14] OUSD, *Department of Defense Financial Management Regulation (DoD FMR)*, Vol. 11A, "Reimbursable Operations Policy," Chapter 6, Appendix I, Washington, D.C., DoD 7000.14-R, updated May 2021a.

[15] As noted in AFI 65-503 Table A19-1, "Military Annual Standard Composite Pay," the standard composite pay rates "do not provide for the portion of military personnel benefits financed by other appropriations, such as the cost of government-furnished quarters for personnel residing in family housing or dormitories; the cost of mess attendant contracts for personnel subsisting in military dining facilities; and commissary and exchange benefits subsidized by appropriated funds" (DAF, 2018).

Table 2.1 provides a summary of the average officer and enlisted standard composite pay rates and their components in FY 2021.

Table 2.1. Summary of Active Duty Officer and Enlisted Standard Composite Pay Rates (FY 2021)

	Average Officer Composite Rate		Average Enlisted Composite Rate	
	Dollars	Percent	Dollars	Percent
Basic Pay	$84,831	48%	$38,048	44%
Retired Pay Accrual	$29,417	17%	$13,238	15%
BAH	$24,986	14%	$15,280	18%
Medicare-Eligible Heath Care Accrual	$4,911	3%	$4,911	6%
BAS	$3,133	2%	$4,500	5%
Special and Incentive Pay	$10,890	6%	$1,635	2%
PCS	$6,012	3%	$2,898	3%
Miscellaneous	$11,241	6%	$5,964	7%
Total	$175,421	100%	$86,474	100%

SOURCE: DAF, 2018, Table A19-2, "Active Air Force Standard Composite Rates by Grade Description."
NOTE: Due to rounding, percentages might not sum to total.

The standard composite pay rates are further broken down by grade in AFI 65-503 Table A19-1, "Military Annual Standard Composite Pay." Figure 2.1 provides a visual summary of how the standard composite pay rates and their components vary by grade.[16]

In many cases, the compensation received by individuals in a composite category will vary because of additional factors such as YOS, location, and career field. For our purposes of estimating MILPERS costs under different future scenarios, we allow standard composite pay rates to vary by year, grade, and YOS in each grade. We discuss each component of the standard composite pay rates below.

---

[16] To inflate the standard composite pay rates to future FY dollars, SAF/FMB guidance is to assume a 2.6 percent annual rate of growth. The average cost of an O-10 is slightly less than the cost of an O-9 because of the difference in BAH. Most O-10s live in basing housing, which is excluded from standard composite pay rates.

Figure 2.1. Standard Composite Pay Rates by Grade (FY 2021)

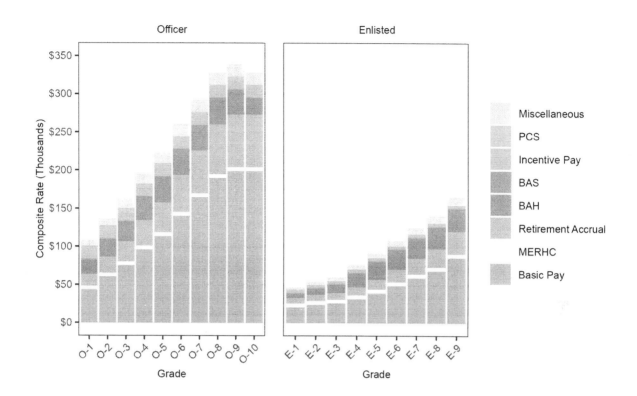

SOURCE: DAF, 2018, Table A19-2, "Active Air Force Standard Composite Rates by Grade Description."
NOTE: MERHC = Medicare-Eligible Retiree Health Care Accrual.

Basic Pay

Basic pay is updated annually and takes effect on January 1 of each year. Basic pay varies by a servicemember's grade and YOS. In FY 2021, basic pay made up 44 percent and 48 percent of standard composite pay rates for enlisted personnel and officers, respectively. It is the largest element of standard composite pay rates by a wide margin.

Year-to-year changes in basic pay are tied to the U.S. Department of Labor's Bureau of Labor Statistics Employment Cost Index (ECI), which measures private sector wage growth.[17] From calendar year 2000 to 2006, the annual change in basic pay equaled the percent change in the ECI plus one-half of one percent. However, Congress might enact basic pay increases below those dictated by the ECI, as it did in calendar years 2012, 2013, and 2014.

---

[17] "Defense Primer: Military Pay Raise," Washington, D.C.: Congressional Research Service, updated December 27, 2021.

Retired Pay Accrual

The U.S. Department of Defense (DoD) has established a trust fund for which money is set aside to pay the expected future retirement benefits of current servicemembers.[18] RPA is tied directly to a servicemember's basic pay, which we allow to vary in our analysis by grade and YOS.[19] In FY 2021, RPA made up 15 percent and 17 percent of standard composite pay rates for officers and enlisted personnel, respectively.

Basic Allowance for Housing

There is not enough government housing for all military members and their families. As a result, the government provides a BAH to help servicemembers obtain commercial housing. Adjustments to a servicemember's BAH might be made to address special situations.

The amount of BAH received will depend on a servicemember's location, pay grade, and whether they have dependents.[20] The rates are set based on a survey of rental properties in geographic locations. Servicemembers stationed in high-cost regions will receive a greater BAH allowance than those stationed in low-cost regions. However, within the standard composite pay rates calculations developed by SAF/FMB, the BAH simply varies by grade. We use SAF/FMB's estimates of BAH by grade in our analysis. In FY 2021, BAH made up 14 and 18 percent of standard composite pay rates for officers and enlisted personnel, respectively.

Medicare-Eligible Retiree Health Care Accrual

MERHC accruals component of standard composite pay rates totaled $4,911 per person in FY 2021; the MERHC is the same for all active duty personnel. Although the MERHC is paid for out of the General Fund of the U.S. Department of the Treasury rather than the MILPERS budget, it is considered part of the military departments' discretionary budget authority and is included in DAF standard composite pay rates.[21] In FY 2021, MERHC made up 3 percent and 6 percent of standard composite pay rates for officers and enlisted personnel, respectively.

Basic Allowance for Subsistence

DoD provides an allowance to offset costs associated with a servicemember's meals.[22] The BAS will vary for officers and enlisted members. Changes over time to the BAS are tied to the

---

[18] James Hosek, Beth J. Asch, and Michael G. Mattock, *Toward Efficient Military Retirement Accrual Charges*, Santa Monica, Calif.: RAND Corporation, RR-1373-A, 2017.

[19] In FY 2021, RPA represented approximately 35 percent of a servicemember's basic pay. RPA does not apply to cadets.

[20] The Defense Travel Management Office publishes the BAH rates; see Defense Travel Management Office, "Allowances," webpage, undated.

[21] The MERHC began being paid out of the General Fund of the U.S. Department of the Treasury in FY 2006. Reimbursement of the MERHC accrual is to be deposited into the miscellaneous receipt account 3041.

[22] The Defense Travel Management Office publishes BAS rates; see Defense Travel Management Office, undated. BAS is not intended to cover the cost of food for family members.

U.S. Department of Agriculture's food cost index. In FY 2021, BAS made up 2 percent and 5 percent of standard composite pay rates for officers and enlisted personnel, respectively.

Special and Incentive Pay

Special and incentive pay can be used to address recruiting and retention needs by increasing compensation in key occupation specialties or critical skill areas (e.g., physicians, dentists, and nurses) and for onerous or hazardous duty assignments or conditions (e.g., aircrews, hazardous duty, hostile fire or imminent danger, and duty at certain places). In addition, special and incentive pay can be used to provide incentives for servicemembers to develop certain skills that are important to national security objectives.[23]

Although a servicemember's special and incentive pay will vary depending on a variety of these factors, within the standard composite pay rates developed by SAF/FMB, special and incentive pay varies only between officer and enlisted personnel. In FY 2021, special and incentive pay made up 6 percent and 2 percent of standard composite pay rates for officers and enlisted personnel, respectively.

Permanent Change of Station

Military services routinely move personnel to assignments at new locations to meet national security requirements and institutional needs.[24] DoD compensates personnel for moving expenses associated with new duty assignments. Compensation for PCS costs will vary by individual and per move based on a variety of factors (e.g., rank, number of dependents, and location of next duty station). Allowable expenses include those related to shipment of household goods and privately owned vehicles, commercial air travel, and temporary lodging. Officers have higher allowances in certain categories (e.g., household goods shipments, travel expenses, and dislocation allowances) and experience shorter tour lengths (also called *time on station*) than enlisted personnel, resulting in more-frequent changes of station.[25]

The PCS component of standard composite pay rates is calculated by dividing the officer and enlisted worldwide PCS costs for a given year by the respective officer and enlisted work years in that year. In FY 2021, PCS made up 3 percent of standard composite rates for both officers and enlisted personnel.

---

[23] DoD, "Military Compensation—Special and Incentive Pay," website, undated.

[24] PCS moves are distinct from deployments or temporary duty travel and are assigned to one of the following six categories: accession travel, separation travel, operational travel, rotational travel, training travel, and organized unit travel (U.S. Government Accountability Office, "Military Compensation: DoD Needs More Complete and Consistent Data to Assess the Costs of Policies on Relocating Personnel," Washington, D.C., GAO-15-173, September 2015).

[25] GAO, 2015.

Miscellaneous

The miscellaneous component of standard composite pay rates covers a variety of costs including family separation allowance, separation payments, social security tax (employer's contribution), overseas station allowances, death gratuities, reenlistment bonuses, special duty assignment pay, clothing allowances, unemployment compensation, and personal money allowances for the O-9 and O-10 pay grades. The miscellaneous component of the standard composite rates tends to increase with grade, although there are some exceptions among officer grades.[26] In FY 2021, miscellaneous costs made up 6 percent and 7 percent of standard composite pay rates for officers and enlisted personnel, respectively.

End Strength and Work Years

Individuals are accessed, promoted between grades, and separated at various points in an FY. As individuals move throughout their military career, their compensation and the applicable standard composite pay rates applied for costing purposes will change. To calculate the MILPERS spending in an FY, in addition to understanding the standard composite pay rates that apply, it is essential to estimate the work years associated with personnel in each grade and at different YOS. Work years represent the amount of time (measured in years-equivalents) that accrues in an FY to individuals in a particular grade. For our purposes, we further break down work years by YOS in each grade.

Historically, work years in an FY were found to correspond closely with the end strength levels calculated at the beginning and end of the FY. For the purposes of our calculations, we use average strength as a proxy for work years.

Estimating Total MILPERS Costs

The standard composite pay rates discussed above provide an estimate of the average cost (per person-year) of active duty military personnel broken out by grade in each FY. As we note, we further allow the basic pay and retirement accrual portion of standard composite pay rates to vary by YOS in each grade. To estimate the total MILPERS costs, we multiply the standard composite pay rates by the work years performed by personnel in each grade and YOS category over the course of the FY and sum the totals.

Personnel Costs Not Included in MILPERS

The standard composite rates and the MILPERS cost accounting approach used by SAF/FMB—and applied in this report—provide a partial view of personnel costs, amounting to

---

[26] In FY 2021, O-8 and O-10 were associated with a somewhat lower miscellaneous cost than O-7 and O-9, respectively.

about 80 percent of direct personnel costs.[27] It is difficult to account for the remaining 20 percent of all direct personnel costs because of the disaggregated structure across the budget. These costs appear in other accounts, including Operation and Maintenance (e.g., the Defense Health Program, DoD Education Activity, and other in-kind benefits), Family Housing, Military Construction (e.g., the Defense Health Agency and DoD Education Activity), Revolving Management Funds (e.g., Defense Commissaries), and Procurement (e.g., DoD Education Activity).

The standard composite pay rates and the MILPERS cost accounting approach also do not consider the indirect costs, or *burdened labor rate*, of military personnel. These include the costs of administrative and other personnel needed to support individuals to perform DAF missions. The cost of personnel to provide such services as training, career field management, and health care fall under MILPERS; however, the demand for these services changes with the size and makeup of the workforce.

The Full Cost of Manpower (FCoM) model is an alternative approach, created by the Office of Cost Assessment and Program Evaluation in the Office of the Secretary of Defense, to estimate personnel costs consistently across DoD.[28] FCoM accounts for cost factors beyond pay, allowances, and fringe benefits, including health care costs and the costs to recruit and train personnel.[29]

From an enterprise perspective, it is important to understand the magnitude of personnel costs not included in MILPERS and how they will change under future manpower and personnel policies. It was beyond the scope of this study to formally account for these costs. Future research might link these costs to manpower and personnel policies and capture them in the tool discussed in Chapter 5.

Conceptual Link Between Policy Options and MILPERS Components

Table 2.2 maps various manpower and personnel policy options to elements of the standard composite pay rates that they affect. White cells denote a decrease in the average cost of that component per airman, black cells denote an increase in the average cost, and gray cells denote no change.

---

[27] Seamus P. Daniels, "Assessing Trends in Military Personnel Costs," Center for Strategic and International Studies, September 2021.

[28] Office of the Secretary of Defense, Office of Cost Assessment and Program Evaluation, *FCOM Military Rates 2017: White Paper—References, Calculations, and Assumptions*, Washington, D.C.: U.S. Department of Defense, April 2017.

[29] Although we used active DAF standard composite pay rates to maintain parity with how DAF MILPERS costs are estimated, we included an additional category—cost to recruit and train new accessions—based on FCoM.

Table 2.2. Relationship Between Manpower and Personnel Policies and Elements of Standard Composite Pay Rates

	Basic (45%)	RPA (16%)	BAH (17%)	MERHC (5%)	BAS (4%)	Special (3%)	PCS (3%)	Misc. (7%)
Reduce end strength								
Shift to more-junior grade mix	↓	↓	↓					↓
Decrease officer-to-enlisted ratio	↓	↓	↓		↑	↓	↓	↓
Cap pay increases below changes in ECI	↓	↓						
Link BAS and/or BAH to other economic indexes			↓		↓			
Reduce percentage of enlisted entering at advanced rank (E-2 or E-3)	↓	↓	↓					↓
Reduce HYT and selective continuation	↓	↓						
Shift competitive officer promotion windows	↓	↓						
Increase assignment durations							↓	
Reduce overseas presence			↓			↓	↓	↓

NOTE: White cells denote a decrease in the average cost of the component per airman. Black cells denote an increase in the average cost of the component per airman. Gray cells denote no change in the average cost of the component per airman.

The policy and cost element combinations in Table 2.2 suggest the following options:

- *Reduce end strength.* End strength reductions limit future MILPERS spending but do not necessarily affect the average cost of an airman. In fact, if the size of the enlisted workforce falls by more than the size of the officer workforce, the average cost of an airman might increase.
- *Shift to a more-junior grade mix.* Basic pay, BAH, and miscellaneous costs are all higher for individuals who are more senior in rank. Shifting to a more-junior grade mix would reduce the costs of these elements.
- *Decrease officer-to-enlisted ratio.* Basic pay and other cost elements are higher for officers than for enlisted personnel. Decreasing the ratio of officers to enlisted personnel would reduce the costs of these elements. Of note, BAS is higher for enlisted personnel, and would increase costs in this scenario.
- *Cap pay increases below changes in ECI.*[30] This would limit growth in basic pay and retired pay accrual. However, increases in special and incentive pay and miscellaneous costs to offset potential negative recruiting and retention effects might reduce savings.
- *Link BAS and/or BAH to other economic indexes.* The Congressional Budget Office (CBO) explored the effects of adjusting all elements of cash pay, including BAS and BAH,[31] based on ECI and an alternative wage index. If ECI grows more slowly than

[30] CBO, *Approaches to Changing Military Compensation*, Washington, D.C., January 2020; DoD, *Report of the Thirteenth Quadrennial Review of Military Compensation*, Washington, D.C., December 2020.

[31] CBO, *Alternative Approaches to Adjusting Military Cash Pay*, September 2021.

other economic indexes used to adjust BAS and BAH, this would reduce the costs of these elements.

- *Reduce the percentage of enlisted personnel entering at advanced rank.* A high percentage of enlisted personnel enter the workforce at the advanced ranks of E-2 or E-3, either because of the length of the initial service commitment or the prior completion of college credits or other approved training experiences.[32] By reducing this percentage, the DAF could increase the share of enlisted personnel at the ranks of E-1 and E-2, which would have effects like those of shifting to a more-junior grade mix.
- *Reduce HYT and selective continuation.* HYT is the maximum number of years an enlisted servicemember can serve at a certain grade; *selective continuation* is the deferment of involuntary retirement or discharge of officers who are not selected for promotion.[33] By reducing HYT and the use of selective continuation, the DAF could decrease overall YOS, which would reduce basic pay and retired pay accrual.
- *Shift competitive officer promotion windows.* By reducing minimum time in grade (TIG) and YOS for individuals to be considered for promotion, the DAF could accelerate promotion timing. This would reduce average YOS of individuals serving at more-senior grades, which would reduce basic pay and retired pay accrual. This would also increase competitiveness of promotions, which, when combined with limited use of selective continuation, would decrease overall YOS.
- *Increase assignment durations.*[34] By increasing assignment durations, the DAF could reduce the number of PCSs and the associated costs.
- *Reduce overseas presence.* Servicemembers stationed overseas are entitled to additional BAH, along with special and incentive pay and miscellaneous pay. PCS costs are also greater for international moves. Reducing the number of servicemembers stationed overseas would reduce these cost elements.

Summary

This chapter presented a brief review of the elements that make up standard composite pay rates. This review supports three findings.

1. Basic pay accounts for the largest share of standard composite pay rates. The implication for future MILPERS spending is that options that limit the rate of growth in basic pay, as opposed to other elements of standard composite pay rates, have the greatest potential to reduce the average cost of an airman.
2. The composition of the workforce by grade and YOS affects basic pay and many other elements of standard composite pay rates. The implication for future MILPERS spending is that options that affect these characteristics have the greatest potential to reduce the average cost of an airman. In addition, personnel policies intended to achieve other

---

[32] AFI 36-2032, *Military Recruiting and Accessions*, Washington, D.C., September 27, 2019c.

[33] AFI 36-2606, *Reenlistment and Extension of Enlistment in the United States Air Force*, Washington, D.C., September 20, 2019; AFI 36-2501, *Officer Promotion and Selective Continuation*, Washington, D.C., April 30, 2021.

[34] Diana S. Correll, "Air Force Extends First-Term, Unaccompanied Tours at Some Overseas Duty Stations to 36 Months," *Air Force Times*, February 12, 2021.

human resource management objectives (e.g., increasing retention of talent) might contribute to greater future MILPERS spending.

3. MILPERS provides an incomplete account of personnel costs. Although beyond the scope of this study, it will be important to consider other direct costs that appear outside the MILPERS budget (i.e., the Defense Health Program) when evaluating steps to improve MILPERS affordability. Indirect costs, such as providing services, career field management, and training, should be considered as well.

Chapter 3. Historical and Cross-Service Analysis of MILPERS Spending

To provide context for the analytic tool and simulations presented in Chapter 4 and Chapter 5, here we briefly describe trends in DAF MILPERS spending over time, and we compare factors that contribute to different MILPERS spending levels across the military services in FY 2021.

DAF MILPERS Spending over Time

Between FY 2000 and FY 2021, DAF spending on MILPERS roughly doubled, increasing from approximately $18 billion to $36 billion in then-year dollars (see the top graph in Figure 3.1). The nominal annual rate of growth, however, has varied over time, ranging from 4.8 percent between FY 2000 and FY 2012 to −1.4 percent between FY 2012 and FY 2016. Between FY 2016 and FY 2021, the annual rate of growth in MILPERS spending increased again to 3.8 percent.

Between FY 2000 and FY 2021, DAF end strength declined by approximately 6 percent, from 355,654 to 335,485 active duty military personnel.[35] Thus, the primary driver of growth in the DAF MILPERS spending is the change in the average cost per active duty personnel (see the bottom graph in Figure 3.1), which increased at an average annual rate of 3.5 percent in nominal terms, from approximately $50,000 to $103,000, between FY 2000 and FY 2021. By comparison, over that same 21-year period, prices in the overall economy and civilian pay grew by only 1.9 percent and 2.3 percent per year, respectively.[36]

Appendix A applies a method to isolate the contribution of changes in standard composite pay rates, end strength, and grade mix to growth in MILPERS spending over time. Consistent with Figure 3.1, analysis using the method shows that changes in standard composite pay rates for active duty personnel most closely explain movements in the DAF's MILPERS budget between FY 2000 and FY 2021, while change in end strength was a less-significant driver of the MILPERS budget. The method applied in Appendix A has the added benefit of isolating the

---

[35] The decline between 2000 and 2021 is less if one compares work years; work year declined from 360,226 in 2000 to 349,460 in 2021; see U.S. Air Force, Financial Management and Comptroller, "Air Force President's Budget FY22," May 2021.

[36] Rates of GDP and civilian pay growth were calculated by RAND from Table 5.1 and Table 5.5 of the FY 2022 DoD Green Book (OUSD, 2021c). Daniels (2021) investigated the causes of military personnel cost growth over this time horizon and earlier periods for all of DoD. Causes of cost growth identified by Daniels include growth in military basic pay above the rate of growth observed in private sector wages, as well as increases in basic allowances for housing and the establishment of Medicare-Eligible Health Care fund (or Tricare for Life) in the FY 2001 National Defense Authorization Act (Daniels, 2021).

effect of shifts in grade mix. It shows that changes in the share of regular DAF end strength assigned to different grades between FY 2000 and FY 2021 are not a significant driver of the DAF's MILPERS budget.

Figure 3.1. DAF MILPERS Spending and End Strength over Time

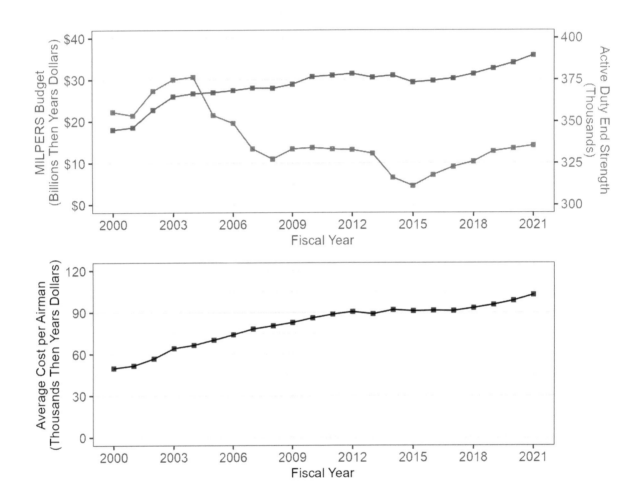

SOURCES: DAF Military Personnel Program President Budget Submissions for FY 2002 to FY 2021.

Comparison of DAF and Other Services' Spending on MILPERS

At approximately $103,000, the average cost of an active duty airman in FY 2021 was comparable with the cost of a Navy sailor ($104,000), and significantly more than the cost of an Army soldier ($98,000) or Marine ($85,000). In this section, we compare end strength, standard composite pay rates, and the grade composition of the military services in FY 2021 to account for these differences.

Table 3.1 summarizes end strength levels for each service, broken out by officer, warrant officer, enlisted personnel, and cadet. In terms of total end strength, the DAF is most like the

Navy. When comparing the workforce compositions, the DAF's active duty end strength is more-heavily weighted toward officers (19.4 percent) relative to the other services (15.5 percent for the Navy, 10.8 percent for the Marines, and 16.3 percent for the Army).[37] Given that basic pay is higher for officers than for enlisted personnel, this partially accounts for the greater average cost of an airman relative to an Army soldier or a Marine.

Table 3.1. Active Duty Military Personnel End Strength Across Services (FY 2021)

	DAF		Navy		Army		Marine Corps	
	Count	Percent	Count	Percent	Count	Percent	Count	Percent
Officer	64,936	19.4%	54,128	15.5%	79,149	16.3%	19,540	10.8%
Warrant Officer	-	0.0%	1,968	0.6%	14,597	3.0%	2,271	1.3%
Enlisted Personnel	266,451	79.4%	287,772	82.6%	387,752	79.8%	159,393	88.0%
Cadet	4,098	1.2%	4,491	1.3%	4,502	0.9%	-	0.0%
Total	**335,485**	**100.0%**	**348,359**	**100.0%**	**486,000**	**100.0%**	**181,204**	**100.0%**

SOURCE: Each service's FY 2022 Military Personnel Program President Budget Submission. End strength values represent end strength on September 30, 2021, and differs slightly from work year estimates for FY 2021.
NOTES: Marine Corps cadets are included in the total for Navy cadets. Due to rounding, percentages might not sum to total.

The average standard composite pay rates for officers, warrant officers, enlisted personnel, and cadets also differ across the services (see Figure 3.2). The DAF's overall average cost per active duty servicemember corresponds most closely with the Navy's but exceeds those of the Army and the Marine Corps. This finding holds for both officers and enlisted personnel.

---

[37] Additionally of note, all other services use warrant officers, whereas the DAF does not.

Figure 3.2. Average Cost of Active Duty Military Personnel Across Services (FY 2021)

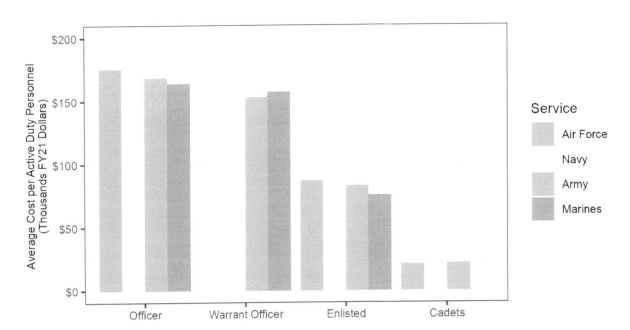

SOURCES: Derived by RAND from Office of the Secretary of Defense (Comptroller), "FY 2021 Department of Defense Military Personnel Composite Standard Pay and Reimbursement Rates," May 11, 2020, and each service's FY 2022 MILPERS President Budget Submission.

The differences in standard composite pay rates across the services for active duty officers and enlisted personnel are partially attributable to differences in grade mix. Figure 3.3 compares the share of officers and enlisted personnel in different grades by service. The DAF grade mix for officers is more senior relative to other services; the DAF's grade mix for enlisted personnel is like that of the Navy and the Army. Marine Corps enlisted personnel are concentrated in more-junior grades relative to the other military services. Given that the rate of basic pay is higher for more-senior individuals, this partially accounts for the greater standard composite pay rates and average cost of a DAF servicemember relative to the Army and the Marines.

Figure 3.3. Officer and Enlisted Grade Composition Across Services (FY 2021)

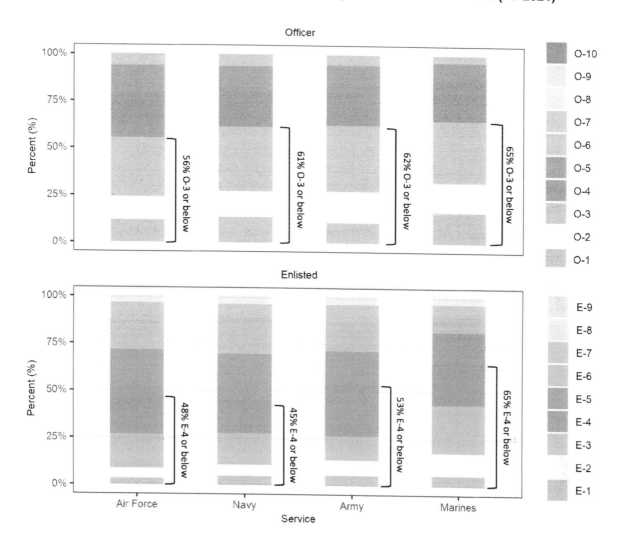

SOURCES: Each service's FY 2022 MILPERS President Budget Submission. Percentages are based on end strength at the end of FY 2021 (September 30, 2021).

Summary

This chapter presented a short analysis of DAF MILPERS spending over time, followed by a comparison of MILPERS spending across the services in FY 2021. The results of this analysis support two findings:

1. Changes in standard composite pay rates for active duty personnel most closely explain movements in the DAF's MILPERS budget from FY 2000 to FY 2021; changes in end strength are the second most-important factor. The implication is that solution options that slow the rate of growth in standard composite pay rates are likely to have the greatest impact on future MILPERS spending.

2. The DAF and Navy have similar active duty standard composite pay rates, whereas the Army and Marines have lower rates. These differences relate to the higher officer-to-

enlisted ratio in the DAF and the more-senior officer and enlisted grade mix in the DAF. The implication is that solutions that decrease the officer-to-enlisted ratio or that produce a more-junior workforce could reduce the average cost of an airman and reduce the MILPERS budget.

Chapter 4. Analytic Framework for Evaluating Implications of MILPERS Affordability Efforts

In Chapter 1, we presented a conceptual framework for examining the savings and implications of different approaches to limiting the growth of the MILPERS budget (Figure 1.1). The framework includes various dependencies:

- Authorized workforce structure and personnel policies influence the makeup of the personnel inventory.
- The makeup of the personnel inventory influences the average cost of an airman.
- The size of the personnel inventory and the average cost of an airman can be used to approximate the total MILPERS budget.

In this chapter, we present an analytic tool implemented in Microsoft Excel that combines components of the conceptual framework. The analytic tool leverages previous models designed to simulate officer promotion and retention.[38] Table 4.1 summarizes the components and data sources used in the analytic tool. After describing the analytic tool, we summarize the classes of workforce and personnel policies that the tool can be applied to.

---

[38] Walsh et al., 2021.

Table 4.1. Elements Contained in Analytic Tool

Component	Description	Data Sources
Authorized workforce structure	• The number of individuals authorized for service by grade and AFSC	• FY 2024 funded authorizations retrieved from MPES in December 2021
Personnel policies	• Accession, promotion, reenlistment, and continuation policies	• AFI 36-2032, *Military Recruiting and Accessions* • AFI 36-2501, *Officer Promotion and Selective Continuation* • AFI 36-2502, *Enlisted Airman Promotion and Demotion Programs*[a] • AFI 36-2606, *Reenlistment and Extension of Enlistment in the United States Air Force*
Personnel inventory	• Structural models of personnel policies • Statistical models of outcomes (e.g., promotion and separation)	• AFIs • Military personnel data system
Average cost of an airman	• Average cost of an airman by grade and YOS	• AFI 65-503, Table A19-2, "Active Air Force Standard Composite Rates by Grade" • 2021 military active and reserve component pay tables
Training and recruiting cost	• Accession, basic skills, advanced training, and recruiting costs	• FY 2021 defense budget materials • FY 2021 military personnel budget
Other workforce categories	• Civilian, contractor, and Air Force Reserves	• FY 2021 SAF/FM Cost of an Airman Tool

NOTE: MPES = Manpower Programming and Execution System; SAF/FM = Assistant Secretary of the Air Force for Financial Management and Comptroller.
[a] AFI 36-2502, *Enlisted Airman Promotion and Demotion Programs*, Washington, D.C., September 27, 2019b.

Components of the Analytic Tool

Authorized Workforce Structure

We treated authorized workforce structure as FY 2024 funded authorizations contained in MPES. Figure 4.1 shows funded authorizations by grade.[39] In total, funded authorizations include 63,376 officers and 268,158 enlisted personnel. Historically, end strength and grade strength have deviated from these figures because of programming processes and external factors. For example, during FY 2021, end strength exceeded funded authorizations, about 30 percent more enlisted personnel were at the ranks of E-1 to E-3 than authorized, and about 40 percent more officers were at the ranks of O-1 to O-2 than authorized.

Of note, the funded authorizations do not distinguish between the grades of O-1 to O-2 or O-7 to O-10 for officers and E-1 to E-3 for enlisted personnel. In addition to grade, each billet has attributes related to occupation (e.g., AFSC), organization (e.g., unit, wing, group, and

---

[39] Because the U.S. Space Force military personnel are budgeted and paid out of the DAF military personnel budget, funded authorizations include these personnel.

command), location (e.g., installation), command level (e.g., wing and above), mission (e.g., tooth versus tail), among other things.

Figure 4.1. Officer and Enlisted Funded Authorizations (FY 2024)

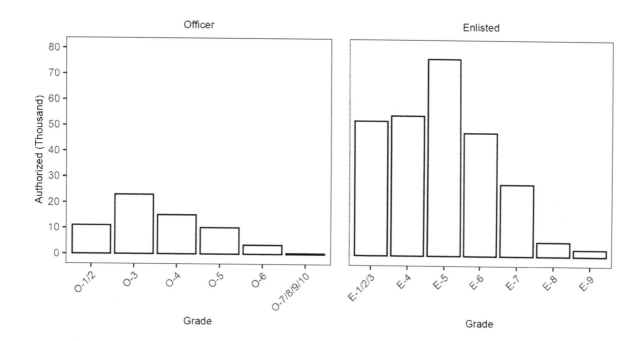

NOTE: MPES extract generated in December 2021.

The analytic tool allows the user to directly change authorizations and grade mixes. The tool also allows the user to apply targeted changes to workforce segments based on billet attributes—for example, applying a 10 percent reduction to positions at the wing level and above.

Personnel Policies

Figure 4.2 gives a high-level view of elements of career field management that shape the workforce (i.e., average strength by grade). Officer promotion, separation, and retirement follow guidelines established by the Defense Officer Personnel Management Act and prescribed in AFIs.[40] Congress grants the services greater latitude to manage the enlisted workforce. Guidelines for enlisted management are also prescribed in various AFIs.[41]

---

[40] DAF, 2021.

[41] DAF, 2019a; DAF, 2019b.

Figure 4.2. High-Level View of Elements of DAF Career Field Management

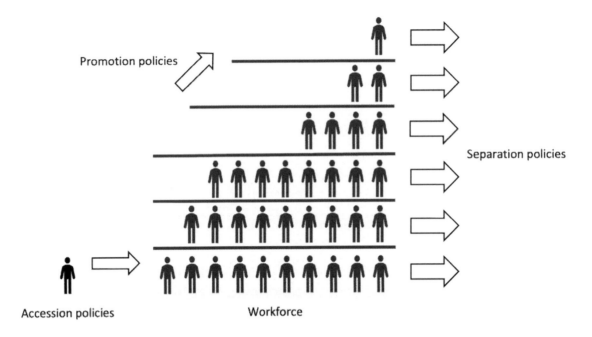

Closed Personnel System

The DAF has a closed personnel system, meaning officers and enlisted personnel enter the workforce at low grades, and positions at higher grades are filled by internal promotion. Officers might enter above the grade of O-1, but this is most prevalent in medical and professional career fields. Many enlisted personnel enter above the grade of E-1, based on the length of the initial service commitment, completion of college semester hours, or completion of training programs, such as Junior Reserve Officers' Training Corps.[42]

Personnel Pyramid

Grade structures for officers and enlisted personnel are somewhat pyramid shaped; there are fewer authorizations at higher ranks. The Defense Officer Personnel Management Act sets the maximum percentage of active duty officers in field-grade ranks (O-4 to O-6) and above for each service. Congress also legislates the maximum percentage of the enlisted workforce that can serve in grades E-8 (2.5 percent) and E-9 (1.25 percent).[43]

Competitive Promotions

Officers and enlisted personnel advance through competitive promotions. Officers become eligible for promotion after meeting minimum TIG requirements. All officers who are fully qualified and meet minimum TIG requirements are selected and promoted to O-2 and O-3. The

---

[42] DAF, 2019c.

[43] "Defense Primer: Military Enlisted Personnel," Washington, D.C.: Congressional Research Service, updated December 1, 2021.

26

number of officers selected for promotion to O-4 to O-6 annually depends on the projected number of vacancies at the next highest grade. Central promotion boards evaluate the records of all officers who are eligible for promotion to O-4 to O-6.[44] Officers are then selected based on order of merit. Most officers selected for promotion to O-4 to O-6 are chosen in the promotion zone (IPZ)—that is, during a one-year window determined by TIG. Promotion processes for general officers differ, but also entail competitive selection.

Enlisted personnel become eligible for promotion after meeting minimum TIG and YOS requirements. All enlisted personnel who are fully qualified and meet minimum TIG and YOS requirements are selected and promoted to E-2, E-3, and E-4.[45] Enlisted personnel eligible for promotion to E-5 and E-6 take the Weighted Airman Promotion System test. A promotion quota is applied in each AFSC, and individuals with the highest scores are selected. Enlisted personnel eligible for promotion to E-7 through E-9 compete and are selected by promotion boards for a limited number of positions in their AFSCs.[46] Unlike officers, a sizable number of enlisted personnel are selected for promotion to E-5 through E-9 across multiyear windows, and not predominantly when they are first considered.

Up-or-Out Career Flow

Because of the competitive nature of selections, some officers and enlisted personnel are not selected for promotion. Officers twice passed for promotion to the grades of O-4 to O-6 might be forced to separate or retire. Involuntary separation or retirement could be deferred for officers chosen for selective continuation.[47] Enlisted personnel reaching HYT might also be forced to separate or retire. HYT has changed over time, and HYT rules might be waived.[48]

Personnel Inventory Projection

We implemented inventory projection models based on the personnel policies described in the previous section. The model simulates the long-run (or steady-state) effects of various policy options. The models follow the logic shown in Figure 4.3. The number of individuals at a given rank and YOS is determined by the sources of flow into that inventory bin. To give a hypothetical example, the number of O-4 with 10 YOS is set based on (1) the number of O-3

---

[44] Career fields are grouped into developmental categories. Competition for promotion is between officers in the same developmental category.

[45] Some individuals might advance to E-4 early through the below-the-zone promotion program.

[46] DAF, 2019b.

[47] DAF, 2021.

[48] DAF, 2019a.

with 9 YOS *retained* and *selected* for promotion to O-4;[49] and (2) the number of O-4 with 9 YOS *retained* and *not selected* for promotion to O-5.[50]

Figure 4.3. Inventory Projection Model

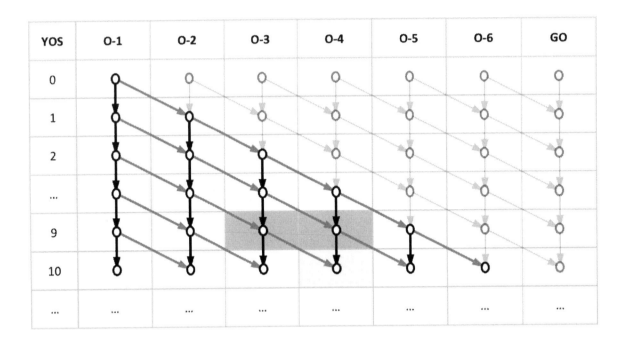

NOTE: GO = general officer.

The same logic applies to all bins in the inventory. In this way, the model computes the probability that an individual entering the workforce will reach any inventory bin. The model also computes the long-run number of individuals in each bin.

Summing down the columns in Figure 4.3 gives the *projected workforce*—that is, the number of individuals by grade. Summing across the rows in Figure 4.3 and taking a weighted average gives the *projected experience level*—that is, the average YOS. Finally, overlaying the military pay table on the rows and columns of Figure 4.3 gives the *projected cost of an airman*—the sum of all cost elements contained in the standard composite pay rates, weighted by the number of individuals by grade and YOS.

The analytic tool applies the inventory model to all enlisted career field groups and to all Line of the Air Force officer categories. Medical and professional officer categories are somewhat different, so the personnel policies that the user selects are not applied to these categories. However, the user could change the end strength of these categories.

---

[49] This equals $Number_{O3,9YOS} \times P(Retained|O3, 9YOS) \times P(Promoted|O3, 9YOS)$.

[50] This equals $Number_{O4,9YOS} \times P(Retained|O4, 9YOS) \times \big(1 - P(Promoted|O4, 9YOS)\big)$.

Accessions

The inventory projection model requires as input the percentage of individuals entering the workforce at each grade. Because the inventory model is applied only to Line of the Air Force officer categories, 100 percent officer accessions are at O-1. By default, the analytic tool sets 77 percent of enlisted personnel as entering at the rank of E-3 and the remaining 23 percent as entering at the rank of E-1. These values reflect the historical percentage of enlisted personnel with a six-year initial service commitment or other experiences counting toward an advanced entry grade.[51]

The analytic tool allows the user to change the percentage of enlisted personnel entering the workforce at the rank of E-3. Using all inputs, the analytic tool automatically determines the total number of annual accessions needed to sustain the workforce (i.e., to reconcile end strength with the authorized workforce structure).

Separations

The inventory projection model requires as input annual separation probabilities. Figure 4.4 shows cumulative continuation rates computed for officers and enlisted personnel by career field grouping.[52] From these, we derived and applied different annual separation probabilities by career field group and YOS. For example, annual separation rates are lowest for officers in operations career fields over the first 10 YOS. Annual separation rates for enlisted personnel increase at 4 YOS and 6 YOS, corresponding to the end of the first enlisted term. Annual separation rates for officers and enlisted personnel plateau from about 15 YOS to 20 YOS and increase after servicemembers become eligible to retire.

The analytic tool allows the user to change HYT for enlisted personnel and to set an annual separation rate for individuals exceeding HYT. The analytic tool also allows the user to set maximum YOS in certain ranks for officers. This allows the user to simulate involuntary separation or retirement for officers who are passed for promotion.[53]

---

[51] Percentages are based on data contained in MILPERS from FY 2017 to FY 2021. In practice, some of these individuals would enter at the rank of E-2, and some would receive the authorized grade upon completing basic military training and initial skills training, as outlined in AFI 36-2032 (DAF, 2019c). To simplify matters, we treated all these cases as entering the workforce at the rank of E-3.

[52] Rates are based on data from FY 2011 to FY 2020.

[53] Historically, separation rates for officers who are not selected IPZ spike during the year after they are passed for promotion. We did not explicitly represent this factor in our statistical models of officer separation.

Figure 4.4. Cumulative Continuation Rates

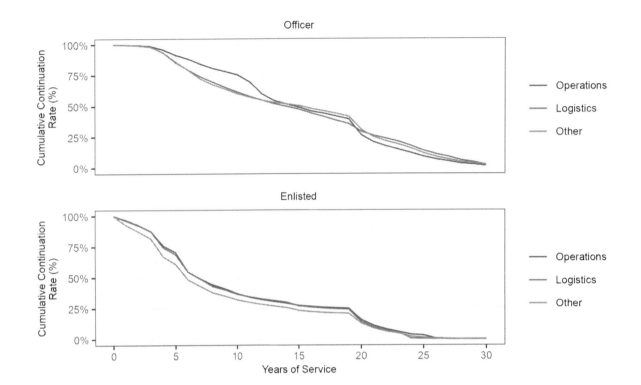

Promotions

The inventory projection model requires as input annual promotion probabilities. We set these to 100 percent for promotion to O-2 and O-3 and E-2 through E-4.

Promotion to all other grades is competitive. The inventory projection model is specified as a nonlinear model. The analytic tool uses the Microsoft Excel Solver add-in to determine promotion rates to minimize differences between projected grade strength and the grade strength corresponding to the authorized workforce structure set by the user. This approximates vacancy-based promotions.

The analytic tool allows the user to adjust the start of promotion windows for officers and enlisted personnel. Given that promotion timing is highly consistent among officers and that most are selected for promotion to field-grade ranks during the one-year windows when they are IPZ, we set the width of promotion windows for officers to one year. Given that promotion timing for enlisted personnel is more variable, the analytic tool allows the user to set multiyear promotion windows for each grade in the enlisted workforce.

Other changes that the user makes to workforce structure or personnel policy indirectly affect promotion rates found by the Solver.

30

Average Cost of an Airman

We computed standard composite pay rates by grade and YOS. To do so, we used FY 2021 values of cost elements contained in AFI 65-503 Table A19-2, "Active Air Force Standard Composite Rates." To compute costs by grade and YOS, we replaced the basic pay and RPA elements with values derived from the FY 2021 military active and reserve component pay tables. We overlaid these table data on the inventory projection to determine the average cost of an airman by grade (averaging across YOS) and the overall average cost of an airman (averaging across grade and YOS).

By changing the authorized workforce structure or personnel policies, the user can indirectly change the makeup of the projected inventory. These changes translate to differences in the average cost of an airman and the total amount of MILPERS spending.

As described in Chapter 2, year-to-year changes in basic pay are linked to changes in ECI. However, Congress might enact basic pay increases below those dictated by the ECI. The analytic tool allows the user to increase or decrease military compensation relative to civilian wages to simulate the future effects of decoupling a change in basic pay from ECI.

Changing military compensation while holding civilian wages constant will produce a retention effect. For example, one study estimated that a 10 percent increase in military basic pay boosts first-term retention by 15 percent to 20 percent and second-term retention by about 10 percent.[54] Using these findings, changes in basic pay in the analytic tool are linked to an elasticity factor that is applied uniformly across all YOS. The elasticity factor specifies the relative percentage increase in annual retention per a 1.0 percent increase in pay. By default, we set the elasticity factor to 1.0. This increases the percentage of officers and enlisted personnel reaching retirement by 18 percent and 21 percent, respectively (i.e., a retirement elasticity factor of 1.18 and 1.21).

Training and Recruiting Costs

To approximate recruiting, accession, basic skills, and advanced training costs, we adopted a modified version of the approach used in the FCoM.[55] We retrieved FY 2021 costs associated with the operation and maintenance training and recruiting budget activity. This contains three subactivities: accession training, basic skills and advanced training, and recruiting and other education and training. For purposes of analysis, we included a subset of the subactivity groups

---

[54] John T. Warner, "The Effect of the Civilian Economy on Recruiting and Retention," in U.S. Department of Defense, *Report of the Eleventh Quadrennial Review of Military Compensation*, supporting research papers, Part 1, Chapter 2, June 2012.

[55] OUSD, *Department of Defense Budget Fiscal Year 2021: Operation and Maintenance Programs (O-1) Revolving and Management Funds (RF-1)*, Washington, D.C., May 2021b.

most relevant to recruiting and training new accessions.[56] Cumulatively, these equaled $293.7 million (training and recruiting), $393.4 million (basic skills and advanced training), and $238.7 million (recruiting and other education and training). These costs exclude flight training ($610.9 million), reflecting an assumption that solutions that affect retention will not be applied, or will be applied along with some compensatory policy, to rated career fields.

In FY 2021, the DAF gained 5,282 officers and 27,228 nonprior service enlisted personnel. We divided the total sum of recruiting and training costs ($930.5 million) by the total number of gains (32,510) to calculate an index for the average cost of recruiting and training a new servicemember ($28,000). Including flight training increases the index ($47,200).

Other Labor Categories

The analytic tool allows the user to make rudimentary changes to labor categories. Specifically, the user can increase the number of civilians, contractors, enlisted personnel, and enlisted officers in the Air Reserve Component (ARC).[57] The standard composite pay rates for these labor categories are $122,000, $200,000, $21,000, and $47,000, respectively.[58] Rates are significantly lower for ARC personnel because they contribute fewer labor days per year.

Studies of replacing military personnel with civilian employees use a variety of conversion factors to equate the efficiency, or work output, of individuals in different workforce categories. These values might range from one civilian replacing one servicemember (1:1) to two civilians replacing three servicemembers (1:1.5).[59] The analytic tool allows the user to change the civilian conversion factor to explore these variations.

Analytic Tool Outputs

Figure 4.5 shows summary outputs produced by the analytic tool. In brief, the solution that generated these outputs involved capping the annual increase in basic pay below the change in ECI, effectively reducing compensation for officers and enlisted personnel.[60] The outputs are

---

[56] In the accession training budget subactivity, we retained SAG 031A (Officer Acquisition), 031B (Recruit Training), and 031D (Reserve Officers Training Corps). In the basic skills and advanced training subactivity, we retained 032A (Specialized Skill Training). Finally, in the recruiting and other education and training subactivity, we retained 033A (Recruiting and Advertising), 033B (Examining), and 033E (Junior ROTC).

[57] ARC standard composite pay rates are based on the weighted average of Air National Guard and Air Force Reserve rates. We did not include rates for full-time ARC personnel. Standard composite pay rates are higher for full-time ARC personnel than for their active duty counterparts and so, from an economic perspective, do not constitute an opportunity for MILPERS savings.

[58] Deputy Chief of Staff for Manpower, Personnel and Services, "Accelerate *Workforce* Change or Lose: An A1 Addendum to CSAF's *Accelerate Change or Lose*," June 2021.

[59] CBO, *Replacing Military Personnel in Support Positions with Civilian Employees*, Washington, D.C., December 2015.

[60] The simulation and results are described in Chapter 5.

displayed for the user-defined solution alongside a baseline solution that does not include any new workforce or personnel policies. The output includes five primary metrics:

- *Total cost.* Separate values are shown for officers, enlisted personnel, nonregular DAF personnel (i.e., contractors, civilians, and ARC), and the cost to recruit and train accessions to sustain the workforce. Reducing compensation reduces the total cost of officers and enlisted personnel. Although the accession costs increase, the solution still leads to a net decrease in total cost.
- *Cost of an airman.* Separate values are shown for officers and enlisted personnel. Reducing compensation reduces the average cost of officers and enlisted personnel.
- *Full-time equivalent (FTE).* Separate values are shown for officers, enlisted personnel, and nonregular DAF personnel. Capping increase in basic pay does not affect work years if the number of annual accessions is set to sustain the workforce.
- *Experience (YOS).* Separate values are shown for officers and enlisted personnel. Reducing compensation reduces retention, producing a more-junior YOS profile. A future version of the tool could display the complete YOS distribution, which is produced by the simulation and used to calculate average YOS.
- *Total accessions needed to sustain the workforce.* Separate values are shown for officers and enlisted personnel. Reducing compensation reduces retention, driving the need for more annual accessions to sustain the workforce.

Figure 4.5. Analytic Tool Quantitative Summary for Solution That Reduces Basic Pay

	Solution	Baseline	Delta	Percent
Total Cost	**$34,228,559,782**	**$35,023,243,541**	**($794,683,759)**	**(-2.3%)**
Officer	$10,555,765,689	$10,732,728,958	($176,963,269)	(-1.6%)
Enlisted	$22,688,265,312	$23,401,820,525	($713,555,213)	(-3%)
Non-RegAF	$0	$0	$0	
Accessions	$984,528,781	$888,694,058	$95,834,724	(10.8%)
Cost of an Airman	**$100,274**	**$102,960**	**($2,686)**	**(-2.6%)**
Officer	$166,560	$169,353	($2,792)	(-1.6%)
Enlisted	$84,608	$87,269	($2,661)	(-3%)
Full Time Equivalent	**331,533**	**331,533**	**0**	**(0%)**
Officer	63,375	63,375	0	(0%)
Enlisted	268,158	268,158	0	(0%)
Non-RegAF	0	0	0	
Experience (YOS)	**6.9**	**7.5**	**-0.5**	**(-7.1%)**
Officer	8.5	8.8	-0.3	(-3.8%)
Enlisted	6.6	7.2	-0.6	(-8.1%)
Total Accessions	**34,545**	**31,182**	**3,363**	**(10.8%)**
Officer	4,652	4,352	299	(6.9%)
Enlisted	29,893	26,830	3,063	(11.4%)

SOURCE: Screenshot from the analytic tool.
NOTE: Non-RegAF = nonregular Air Force personnel.

The analytic tool displays additional summary outputs by grade (i.e., authorized individuals, assigned individuals, average standard composite pay rates, and average YOS).

Solution Space

Table 4.2 summarizes the inputs to the analytic tool that the user can adjust. By doing so, the user can specify, simulate, and observe the outcomes of a wide variety of solution options (see Appendix B).

- *Changes in workforce efficiency.* The user can reduce manpower requirements in different workforce segments, corresponding to increased efficiency because of technology or organization and mission redesign. These changes will directly translate to changes in FTE personnel; the absolute change will depend on the magnitude of the reduction and the size of the affected workforce segment.
- *Changes in workforce design.* The user can reallocate manpower to more-economical labor categories by converting officer requirements to enlisted, converting active duty requirements to reserve, converting military requirements to the civilian or contractor workforce and changing the grade mix.
- *Workforce characteristics.* The user can apply personnel policies to produce a less-experienced, but more-economical workforce, for example, accelerating promotion timing combined with HYT losses, reducing HYT and selective continuation, or reducing entry grades for first-term enlisted personnel.
- *Compensation.* The user can apply changes to compensation to simulate decoupling a change in basic pay from ECI. This is specified as a sustained increase or decrease in pay in then-year dollars.

Table 4.2. Elements Contained in the Analytic Tool

Element	User Input
Authorized workforce structure	• Global changes to end strength and grade strength • Targeted reductions to workforce segments
Personnel policies	• HYT • Start of promotion windows • Percentage enlisting at advanced rank
Average cost of an airman	• Compensation • Pay-related change in retention
Training and recruiting cost	• None
Other workforce categories	• Number of civilians, contractors, ARC enlisted personnel, and ARC officers • FTE personnel conversion factor

Summary

This chapter described an analytic tool that implements a conceptual framework for examining and comparing different approaches to limit the growth of the MILPERS budget. The chapter supports four findings:

1. MILPERS spending can be limited in a variety of ways that do not involve reducing end strength. These include shifting work to more-economical workforce segments and workforce categories, changing the compensation of the workforce, or limiting growth in compensation.
2. Because of the interconnected nature of personnel policies, the personnel inventory, and MILPERS spending, personnel policies could be applied to increase MILPERS affordability. Relatedly, personnel policies designed to achieve nonfinancial objectives, such as retaining skilled servicemembers, will nonetheless affect MILPERS spending.
3. Because of the interconnected nature of the components of the personnel system, solution options entail such trade-offs as reduced end strength, reduced experience levels, and increased recruiting and accession costs.
4. An analytic tool like the one described here can be used to simulate and understand the interplay between these components.

Chapter 5. Application of the Analytic Tool

In this chapter, we apply the analytic tool to a wide variety of workforce and personnel policy options to illustrate how it might be used and to shed light on how the effects of different policies compare.[61] The intent of the tool is to provide order-of-magnitude estimates and to illustrate trade-offs that the options entail. The analytic tool can be used as a starting point to identify promising options for planners to refine.

The eight simulations we present here center on policies that have the potential to reduce MILPERS costs and were chosen based on discussions with the research sponsor. The simulations are (1) reducing end strength; (2) increasing the number of enlisted personnel per officer; (3) modifying officer and enlisted grade structures to include a greater share of junior ranks; (4) limiting annual growth in compensation; (5) reducing enlisted HYT; (6) reducing the percentage of enlisted personnel entering the workforce at advanced rank; (7) accelerating officer promotion timing; and (8) civilianizing officer positions. We model each of these policies in isolation; for example, we vary end strength in the first simulation only. However, the policies could be combined. In addition, we present a single example for each policy, although the size of the change for each policy can be varied.

Simulation 1. Reduce Total End Strength

In the first policy simulation, we reduce end strength without changing grade mix or standard composite pay rates. Specifically, we simulate a reduction in end strength of 6.9 percent, which reflects a reversion to end strength from the recent low point in FY 2015.[62] A reduction could be achieved by a combination of divesting from mission areas (e.g., those with low alignment to the National Defense Strategy and the National Defense Authorization Act), adopting technology to increase workforce efficiency, or undertaking organizational redesign to eliminate redundant positions.[63]

Figure 5.1 shows the analytic tool's top-line summary for this solution option. A 6.9 percent reduction in end strength reduces annual overall spending by 6.9 percent ($2.4 billion). The

---

[61] We use the FY 2024 end strength authorization levels and FY 2021 standard composite pay rates for our baseline. In practice, one could adjust the end strength levels and standard composite pay rates used in the baseline to better reflect future assumptions.

[62] FY 2015 was a recent low point in end strength levels at roughly 311,000. Since then, end strength has risen to 334,000 in FY 2021.

[63] Lance Menthe, Dahlia Anne Goldfeld, Abbie Tingstad, Sherrill Lingel, Edward Geist, Donald Brunk, Amanda Wicker, Sarah Soliman, Balys Gintautas, Anne Stickells, and Amado Cordova, *Technology Innovation and the Future of Air Force Intelligence Analysis: Volume 1, Findings and Recommendations*, Santa Monica, Calif.: RAND Corporation, RR-A341-1, 2022.

itemized costs of officers and enlisted personnel decrease, paralleling the change in the number of FTE personnel. The cost of accessions also decreases, paralleling the change in the number needed to sustain a smaller workforce. The average cost of an airman and average experience are unchanged. All subsequent options reduce MILPERS spending by reducing the average cost of an airman.

Figure 5.1. Effect of a 6.9 Percent Reduction in End Strength

	Solution	Baseline	Delta	Percent
Total Cost	$32,606,639,737	$35,023,243,541	($2,416,603,804)	(-6.9%)
Officer	$9,992,170,660	$10,732,728,958	($740,558,298)	(-6.9%)
Enlisted	$21,787,094,909	$23,401,820,525	($1,614,725,616)	(-6.9%)
Non-RegAF	$0	$0	$0	
Accessions	$827,374,168	$888,694,058	($61,319,890)	(-6.9%)
Cost of an Airman	$102,960	$102,960	$0	(0%)
Officer	$169,353	$169,353	$0	(0%)
Enlisted	$87,269	$87,269	$0	(0%)
Full Time Equivalent	308,657	331,533	-22,876	(-6.9%)
Officer	59,002	63,375	-4,373	(-6.9%)
Enlisted	249,655	268,158	-18,503	(-6.9%)
Non-RegAF	0	0	0	
Experience (YOS)	7.5	7.5	0.0	(0%)
Officer	8.8	8.8	0.0	(0%)
Enlisted	7.2	7.2	0.0	(0%)
Total Accessions	29,031	31,182	-2,152	(-6.9%)
Officer	4,052	4,352	-300	(-6.9%)
Enlisted	24,979	26,830	-1,851	(-6.9%)

SOURCE: Screenshot from the analytic tool.
NOTE: Non-RegAF = nonregular Air Force personnel.

Simulation 2. Increase the Number of Enlisted Personnel per Officer

In the second policy simulation, we converted 20 percent of officer positions (12,675) to enlisted positions. This would cause the number of enlisted personnel per officer to increase from 4.2 to 5.5. For comparison, in FY 2021 the Navy, Marine Corps, and Army had 5.3, 8.3, and 4.9 enlisted personnel per officer, respectively.[64] This could be achieved by converting technical officer positions, such as those in remotely piloted aircraft career fields, to the enlisted workforce.[65]

---

[64] These calculations reflect the end strength values shown in Table 3.1. If we include warrant officers in the officer counts, the enlisted personnel per officer plus warrant officer ratios fall to 5.1, 7.5, and 4.1 for the Navy, Marine Corps, and Army, respectively.

[65] Todd Harrison, "Rethinking the Role of Remotely Crewed Systems in the Future Force," Center for Strategic and International Studies, March 3, 2021.

Figure 5.2 shows the top-line summary for this solution option. Converting 20 percent of officer positions to enlisted positions reduces annual overall spending by 2.9 percent ($1 billion). The average cost of an airman decreases by 3.1 percent ($3,138), reflecting the larger share of enlisted personnel making up the workforce. Total FTE personnel is unchanged; however, a portion of labor shifts from officers to enlisted personnel. Average experience decreases slightly (0.8 percent, or 0.1 YOS) due again to the larger share of enlisted personnel, who tend to separate with fewer YOS. Finally, the number of accessions and associated costs increase because of the larger share of enlisted personnel, who have higher annual separation rates than officers. In addition to considering the changes in cost and experience, the DAF would need to determine whether the savings from this option justify the higher turnover.

Figure 5.2. Effect of Converting 20 Percent of Officer Positions to Enlisted Positions

	Solution	Baseline	Delta	Percent
Total Cost	$33,994,163,236	$35,023,243,541	($1,029,080,305)	(-2.9%)
Officer	$8,586,183,167	$10,732,728,958	($2,146,545,792)	(-20%)
Enlisted	$24,507,952,265	$23,401,820,525	$1,106,131,740	(4.7%)
Non-RegAF	$0	$0	$0	
Accessions	$900,027,804	$888,694,058	$11,333,746	(1.3%)
Cost of an Airman	$99,822	$102,960	($3,138)	(-3%)
Officer	$169,353	$169,353	$0	(0%)
Enlisted	$87,269	$87,269	$0	(0%)
Full Time Equivalent	331,533	331,533	0	(0%)
Officer	50,700	63,375	-12,675	(-20%)
Enlisted	280,833	268,158	12,675	(4.7%)
Non-RegAF	0	0	0	
Experience (YOS)	7.4	7.5	-0.1	(-0.8%)
Officer	8.8	8.8	0.0	(0%)
Enlisted	7.2	7.2	0.0	(0%)
Total Accessions	31,580	31,182	398	(1.3%)
Officer	3,482	4,352	-870	(-20%)
Enlisted	28,098	26,830	1,268	(4.7%)

SOURCE: Screenshot from the analytic tool.

Simulation 3. Modify Grade Structures

In the third policy simulation, we modified the officer and enlisted grade mixes without changing the total numbers of officers or enlisted personnel. We set the grade mixes to match those of the Marine Corps (see Figure 3.3). This amounts to a 2.5 percent increase in the number of officers at the ranks of O-1 through O-3 and a 9.5 percent increase in the number of enlisted personnel at the ranks of E-1 through E-4.

Figure 5.3 shows the top-line summary for this solution option. Shifting to a more-junior grade mix reduces annual overall spending by 4.2 percent ($1.5 billion). The average cost of an airman decreases by 4.5 percent ($4,596), reflecting changes for both officers and enlisted personnel. FTE personnel is unchanged. Average experience drops by 6.6 percent (0.5 YOS) because of the reduced opportunity for individuals to be promoted to higher grades and the ensuing separations. The effect is larger for enlisted personnel because of the more-restrictive values for HYT. Finally, the number of accessions and associated costs increase because of the larger number of individuals separating during each year.

Figure 5.3. Effect of Modifying Officer and Enlisted Grade Structure to Resemble the Marine Corps

	Solution	Baseline	Delta	Percent
Total Cost	$33,550,188,445	$35,023,243,541	($1,473,055,097)	(-4.2%)
Officer	$10,549,650,303	$10,732,728,958	($183,078,656)	(-1.7%)
Enlisted	$22,061,123,204	$23,401,820,525	($1,340,697,322)	(-5.7%)
Non-RegAF	$0	$0	$0	
Accessions	$939,414,938	$888,694,058	$50,720,880	(5.7%)
Cost of an Airman	$98,364	$102,960	($4,596)	(-4.5%)
Officer	$166,464	$169,353	($2,889)	(-1.7%)
Enlisted	$82,269	$87,269	($5,000)	(-5.7%)
Full Time Equivalent	331,533	331,533	0	(0%)
Officer	63,375	63,375	0	(0%)
Enlisted	268,158	268,158	0	(0%)
Non-RegAF	0	0	0	
Experience (YOS)	7.0	7.5	-0.5	(-6.6%)
Officer	8.6	8.8	-0.2	(-2.6%)
Enlisted	6.6	7.2	-0.6	(-7.7%)
Total Accessions	32,962	31,182	1,780	(5.7%)
Officer	4,374	4,352	22	(0.5%)
Enlisted	28,587	26,830	1,758	(6.6%)

SOURCE: Screenshot from the analytic tool.

Simulation 4. Reduce Compensation

The *Report of the Ninth Quadrennial Review of Military Compensation* determined that regular military compensation around the 70th percentile of equivalently educated civilians would allow the military to recruit and retain the quantity and quality of individuals needed; however, current regular military compensation far exceeds this benchmark.[66] The FY 2004 National Defense Authorization Act formally linked basic pay increases to the ECI, but Congress can enact changes below those dictated by the ECI, a flexibility it has exercised in the past. In the fourth policy simulation, we applied a 1.5 percent reduction to the basic pay and RPA elements of standard composite pay rates to simulate a foregone pay increase.

Figure 5.4 shows the top-line summary for this solution option. Reducing compensation reduces annual overall spending by 2.3 percent ($794.7 million). The average cost of an airman decreases by 2.6 percent ($2,686), reflecting changes for both officers and enlisted personnel.

---

[66] DoD, *Report of the Ninth Quadrennial Review of Military Compensation*, Vol. 1, Washington, D.C., March 2002.

FTE personnel is unchanged. Average experience drops by 7.1 percent (0.5 YOS) because of the negative retention effect of reducing pay. This indirectly contributes to the reduced cost of an airman, as YOS is one of the determinants of basic pay and RPA. Finally, the number of accessions and associated costs increase because of reduced retention.[67]

Including flight training in the average cost to recruit and train a new servicemember reduces the annual savings to $731.5 million. The implications are that the full savings from this option would be less than 2.3 percent and that it would be economical for the DAF to replace pay with bonuses in a targeted manner to increase retention in career fields (i.e., rated officers) with high production costs.

Figure 5.4. Effect of Reducing Annual Increase in Basic Pay

	Solution	Baseline	Delta	Percent
Total Cost	$34,228,559,782	$35,023,243,541	($794,683,759)	(-2.3%)
Officer	$10,555,765,689	$10,732,728,958	($176,963,269)	(-1.6%)
Enlisted	$22,688,265,312	$23,401,820,525	($713,555,213)	(-3%)
Non-RegAF	$0	$0	$0	
Accessions	$984,528,781	$888,694,058	$95,834,724	(10.8%)
Cost of an Airman	$100,274	$102,960	($2,686)	(-2.6%)
Officer	$166,560	$169,353	($2,792)	(-1.6%)
Enlisted	$84,608	$87,269	($2,661)	(-3%)
Full Time Equivalent	331,533	331,533	0	(0%)
Officer	63,375	63,375	0	(0%)
Enlisted	268,158	268,158	0	(0%)
Non-RegAF	0	0	0	
Experience (YOS)	6.9	7.5	-0.5	(-7.1%)
Officer	8.5	8.8	-0.3	(-3.8%)
Enlisted	6.6	7.2	-0.6	(-8.1%)
Total Accessions	34,545	31,182	3,363	(10.8%)
Officer	4,652	4,352	299	(6.9%)
Enlisted	29,893	26,830	3,063	(11.4%)

SOURCE: Screenshot from the analytic tool.

Simulation 5. Reduce High Year of Tenure

In February 2019, HYT limits were increased for E-4 (from 8 to 10 YOS), E-5 (from 15 to 20 YOS), and E-6 (from 20 to 22 YOS).[68] Prior to that, HYT values were reduced in 2013 and 2010, after being increased in 2003 and 2001. In the fifth policy simulation, we simulate the effect of reverting HYT limits to previous levels (8, 15, and 20 YOS for E-4, E-5, and E-6, respectively).

---

[67] To determine the sensitivity of results to the elasticity factor, we repeated the simulation after changing the value from 1.0 percent. Setting the elasticity factor to 0.5 percent reduced the level of savings from $794 to $514 million, whereas setting the factor to 1.5 percent boosted the level of savings to $1.1 billion.

[68] Secretary of the Air Force Public Affairs, U.S. Air Force, "Air Force Extends High Year of Tenure for E-4s Through E-6s," Washington, D.C.: U.S. Air Force, October 22, 2018.

Figure 5.5 shows the top-line summary for this solution option. Reverting HYT to 2018 levels reduces annual overall spending by less than 0.1 percent ($6.9 million). The average cost of an airman decreases very slightly because of the small change in the average cost of enlisted personnel ($24). FTE personnel is unchanged. Average experience decreases very slightly because individuals who reach HYT do so sooner and are involuntarily separated with fewer YOS. Finally, the number of accessions and associated costs increase very slightly because of reduced retention.

The savings in this simulation are quite small. Further reducing HYT of E-5 and E-6 to 13 and 18 YOS, respectively, increases annual savings to $45.3 million.

Figure 5.5. Effect of Reducing High Year of Tenure for E-4, E-5, and E-6 to 2018 Levels

	Solution	Baseline	Delta	Percent
Total Cost	$35,016,323,760	$35,023,243,541	($6,919,781)	(0%)
Officer	$10,732,728,958	$10,732,728,958	$0	(0%)
Enlisted	$23,394,013,927	$23,401,820,525	($7,806,598)	(0%)
Non-RegAF	$0	$0	$0	
Accessions	$889,580,875	$888,694,058	$886,817	(0.1%)
Cost of an Airman	$102,936	$102,960	($24)	(0%)
Officer	$169,353	$169,353	$0	(0%)
Enlisted	$87,240	$87,269	($29)	(0%)
Full Time Equivalent	331,533	331,533	0	(0%)
Officer	63,375	63,375	0	(0%)
Enlisted	268,158	268,158	0	(0%)
Non-RegAF	0	0	0	
Experience (YOS)	7.5	7.5	0.0	(-0.2%)
Officer	8.8	8.8	0.0	(0%)
Enlisted	7.2	7.2	0.0	(-0.2%)
Total Accessions	31,213	31,182	31	(0.1%)
Officer	4,352	4,352	0	(0%)
Enlisted	26,861	26,830	31	(0.1%)

SOURCE: Screenshot from the analytic tool.

Simulation 6. Reduce the Percentage of Enlisted Personnel Entering with Advanced Rank

From FY 2016 to FY 2021, 77 percent of enlisted personnel entered the workforce with advanced enlistment rank or accelerated promotion. This reflects a decrease from more than 90 percent in FY 2012. In the sixth policy simulation, we reduce this value to 50 percent. This can be achieved by reducing the number of six-year enlistees or reducing credit for college and participation in other programs. A second-order effect of this change, not captured in our simulation, is the decreased retention after the fourth YOS that is caused by the larger share of four-year enlistees completing their first term.

Figure 5.6 shows the top-line summary for this solution option. Limiting the percentage of enlisted personnel entering with advanced rank reduces annual overall spending by 0.8 percent ($291.9 million). The average cost of enlisted personnel decreases by 1.2 percent ($1,081), reflecting a greater number at the ranks of E-1 and E-2. FTE personnel is unchanged. Average experience of enlisted personnel drops slightly because of the prolonged time for some to reach senior ranks and the gating effects of HYT. Finally, the number of accessions and associated costs increase slightly because of the greater number of enlisted separations.

Figure 5.6. Effect of Reducing Percentage of Enlisted Personnel Entering with Advanced Rank

	Solution	Baseline	Delta	Percent
Total Cost	**$34,731,346,044**	**$35,023,243,541**	**($291,897,497)**	**(-0.8%)**
Officer	$10,732,728,958	$10,732,728,958	$0	(0%)
Enlisted	$23,111,817,972	$23,401,820,525	($290,002,553)	(-1.2%)
Non-RegAF	$0	$0	$0	
Accessions	$886,799,114	$888,694,058	($1,894,944)	(-0.2%)
Cost of an Airman	**$102,085**	**$102,960**	**($875)**	**(-0.8%)**
Officer	$169,353	$169,353	$0	(0%)
Enlisted	$86,187	$87,269	($1,081)	(-1.2%)
Full Time Equivalent	**331,533**	**331,533**	**0**	**(0%)**
Officer	63,375	63,375	0	(0%)
Enlisted	268,158	268,158	0	(0%)
Non-RegAF	0	0	0	
Experience (YOS)	**7.5**	**7.5**	**0.0**	**(0.2%)**
Officer	8.8	8.8	0.0	(0%)
Enlisted	7.2	7.2	0.0	(0.3%)
Total Accessions	**31,116**	**31,182**	**-66**	**(-0.2%)**
Officer	4,352	4,352	0	(0%)
Enlisted	26,763	26,830	-66	(-0.2%)

SOURCE: Screenshot from the analytic tool.

Simulation 7. Shift Competitive Officer Promotion Windows

Title 10 sets minimum TIG requirements for promotion of officers as follows: 18 months for promotion from O-1 to O-2, two years for promotion from O-2 to O-3, three years for promotion to O-4 and O-5, and one year for promotion to O-6 and O-7. Historically, promotion timing for DAF officers has exceeded these minimums. By shifting competitive promotion windows to lower minimum TIGs, the DAF could reduce average YOS for individuals at higher ranks, thereby reducing personnel costs. This would also increase competitiveness of promotions, which, when combined with limited use of selective continuation, would decrease overall YOS. In the seventh policy simulation, we accelerated due-course promotion of officers to the ranks of O-4, O-5, and O-6 to 8, 13, and 19 YOS, respectively. In addition, we limited selective continuation for individuals twice passed for promotion to O-5 to 50 percent.

Figure 5.7 shows the top-line summary for this solution option. Shifting competitive promotion windows reduces annual overall spending by 0.3 percent ($92.3 million).[69] The average cost of an officer decreases by 0.9 percent ($1,550), reflecting the lower YOS of officers at field-grade ranks. FTE personnel is unchanged. Average experience of officers drops somewhat because of the slightly greater number of individuals who were passed for promotion to field-grade ranks and involuntarily separated. Finally, the number of accessions and associated costs increase slightly because of the greater number of officers who separate annually.

Figure 5.7. Effect of Shifting Competitive Officer Promotion Windows

	Solution	Baseline	Delta	Percent
Total Cost	**$34,931,016,257**	**$35,023,243,541**	**($92,227,284)**	**(-0.3%)**
Officer	$10,634,511,972	$10,732,728,958	($98,216,986)	(-0.9%)
Enlisted	$23,401,820,525	$23,401,820,525	$0	(0%)
Non-RegAF	$0	$0	$0	
Accessions	$894,683,760	$888,694,058	$5,989,702	(0.7%)
Cost of an Airman	**$102,663**	**$102,960**	**($296)**	**(-0.3%)**
Officer	$167,803	$169,353	($1,550)	(-0.9%)
Enlisted	$87,269	$87,269	$0	(0%)
Full Time Equivalent	**331,533**	**331,533**	**0**	**(0%)**
Officer	63,375	63,375	0	(0%)
Enlisted	268,158	268,158	0	(0%)
Non-RegAF	0	0	0	
Experience (YOS)	**7.4**	**7.5**	**-0.1**	**(-1%)**
Officer	8.4	8.8	-0.4	(-4.3%)
Enlisted	7.2	7.2	0.0	(0%)
Total Accessions	**31,392**	**31,182**	**210**	**(0.7%)**
Officer	4,563	4,352	210	(4.8%)
Enlisted	26,830	26,830	0	(0%)

SOURCE: Screenshot from the analytic tool.

Simulation 8. Civilianize Officer Acquisition and STEM Positions

The total force comprises over 16 labor cost categories including active, reserve, and civilian personnel. Variations among labor categories have important implications for affordability and mission design.

The total force contains more than 150,000 DAF civilians, and there might be opportunities to expand the use of this labor category.[70] For example, there are 8,131 FY 2024 authorizations for officers in acquisition and science, technology, engineering, and mathematics (STEM) specialties (i.e., 6-series AFSCs). Most officers in these positions do not engage in combat. In

---

[69] Limiting selective continuation for individuals twice passed for promotion to O-5 without accelerating promotion timing produced almost no effect.

[70] U.S. Air Force, "Air Force Demographics," webpage, Air Force's Personnel Center, undated.

addition, the knowledge, skills, and abilities needed for these positions evolve rapidly and are available in commercial markets. For these reasons, a share of officer acquisition and STEM positions might be well suited for civilian personnel. In the final policy simulation, we converted 40 percent of these positions (3,252) to the civilian workforce. Because the grade pyramid among officers in STEM positions resembles the overall officer grade pyramid, a proportional reduction of these positions does not affect the overall officer grade mix.

Figure 5.8 shows the top-line summary for this solution option. Civilianizing 40 percent of officer acquisition and STEM positions reduces annual overall spending by 0.4 percent ($157 million). The average cost of an airman decreases by 0.6 percent ($648), reflecting the larger share of enlisted personnel making up the remaining workforce. Total FTE personnel is unchanged; however, a portion of labor shifts from officers to nonregular DAF (i.e., civilian) personnel. Average experience decreases very slightly among officers. This is because acquisition and STEM billets are made up of a slightly more-senior grade mix. Finally, the cost of officer accessions decreases, paralleling the change in the number needed to sustain a smaller workforce.

Figure 5.8. Effect of Civilianizing Officer Acquisition and STEM Positions

	Solution	Baseline	Delta	Percent
Total Cost	$34,866,290,798	$35,023,243,541	($156,952,744)	(-0.4%)
Officer	$10,185,171,617	$10,732,728,958	($547,557,341)	(-5.1%)
Enlisted	$23,401,820,525	$23,401,820,525	$0	(0%)
Non-RegAF	$396,744,000	$0	$396,744,000	
Accessions	$882,554,656	$888,694,058	($6,139,402)	(-0.7%)
Cost of an Airman	$102,312	$102,960	($648)	(-0.6%)
Officer	$169,407	$169,353	$54	(0%)
Enlisted	$87,269	$87,269	$0	(0%)
Full Time Equivalent	331,533	331,533	0	(0%)
Officer	60,123	63,375	-3,252	(-5.1%)
Enlisted	268,158	268,158	0	(0%)
Non-RegAF	3,252	0	3,252	
Experience (YOS)	7.5	7.5	0.0	(-0.2%)
Officer	8.8	8.8	0.0	(-0.2%)
Enlisted	7.2	7.2	0.0	(0%)
Total Accessions	30,967	31,182	-215	(-0.7%)
Officer	4,137	4,352	-215	(-4.9%)
Enlisted	26,830	26,830	0	(0%)

SOURCE: Screenshot from the analytic tool.

Chapter Summary and Discussion

This chapter detailed eight simulations that involve using the analytic tool to explore workforce and personnel policies to reduce future MILPERS spending. The outcomes from all eight simulations are presented in Table 5.1 and underscore the trade-offs between cost, capability (i.e., FTE), and experience. Given the exponential growth that occurs in costs over time, these costs would compound across multiple years.

Table 5.1. Comparison of Outcomes from Baseline Scenario and Potential Policy Solutions

Simulation	Total Cost (billions)	Cost of an Airman	FTE	Experience (YOS)	Accessions
1. Reduce end strength	$32.6	$102,960	308,657	7.5	29,031
2. Increase number of enlisted personnel per officer	$34.0	$99,822	331,533	7.4	31,580
3. Modify grade mix	$33.6	$98,364	331,533	7.0	32,962
4. Reduce compensation	$34.2	$100,274	331,533	6.9	34,545
5. Reduce enlisted HYT	$35.0	$102,936	331,533	7.5	31,213
6. Reduce enlistment with advanced rank	$34.7	$102,085	331,533	7.5	31,116
7. Shift competitive promotion windows	$34.9	$102,663	331,533	7.4	31,392
8. Civilianize officer and/or STEM specialties acquisition	$34.8	$102,312	331,533	7.5	30,967
Baseline	**$35.0**	**$102,960**	331,533	7.5	31,182

NOTE: Cells in dark red denote a decrease of 2.5 percent or more from baseline. Cells in light red denote a decrease of more than 0.0 percent from baseline. Cells in white denote no change from baseline. Cells in light green denote an increase of more than 0.0 percent from baseline. Cells in dark green denote an increase of 2.5 percent or more from baseline.

The levels of savings that the solutions entail, although modest in relative terms, equate to significant amounts. For example, the annual savings in the second simulation (i.e., increase number of enlisted personnel per officer) equals $1,209.1 million. Given that the average cost of an airman in the second simulation equals $99,822 and the average cost of a civilian equals $122,000, the savings could be repurposed to support 10,300 additional servicemembers or 8,400 additional civilians. Alternatively, the average variable cost of F-16C and F-35A flying hours in FY 2021 equal $10,361 and $17,963, respectively.[71] The savings from the second simulation could be repurposed to support 99,300 additional F-16C flying hours or 57,300 F-35A flying hours. Table 5.2 shows these conversions for the eight solution options.[72]

Cost growth is exponential. As a result, the annual savings shown in Table 5.2 would increase in each subsequent year relative to the baseline scenario. In addition to the overall savings increasing, the differences between the simulations would also grow larger over time.

---

[71] OUSD, *Fiscal Year (FY) 2021 Department of Defense (DoD) Fixed Wing and Helicopter Reimbursement Rates*, Washington, D.C., October 2020.

[72] The conversions of savings are based on the output of the tool; they are not yet included in the tool itself.

Table 5.2. Conversions of Savings from Potential Policy Solutions

Simulation	Savings (millions)	Extra Airmen (thousands)	Extra Civilians (thousands)	Extra F-16C Flying Hours (thousands)	Extra F-35A Flying Hours (thousands)
1. Reduce end strength	$2,416.6	n/a	19.8	233.2	134.5
2. Increase number of enlisted personnel per officer	$1,029.1	10.3	8.4	99.3	57.3
3. Modify grade mix	$1,473.1	15.0	12.1	142.2	82.0
4. Reduce compensation	$794.7	7.9	6.5	76.7	44.2
5. Reduce enlisted HYT	$6.9	0.1	0.1	0.7	0.4
6. Reduce enlistment with advanced rank	$291.9	2.9	2.4	28.2	16.2
7. Shift competitive promotion windows	$92.2	0.9	0.8	8.9	5.1
8. Civilianize officer and/or STEM specialties acquisition	$157.0	1.5	1.3	15.1	8.7

These results support three findings:

1. Reducing end strength produces the lowest total cost, but this option sacrifices capability (FTE). Modifying the grade mix or capping pay increases below ECI also produces significant savings, but these options sacrifice experience (YOS). No solution reduces costs without sacrificing capability or experience, although the option to increase the number of enlisted personnel per officer performs well across this trade space.
2. Certain personnel policies would allow the DAF to save tens of millions to hundreds of millions of dollars annually.
3. Larger levels of savings require changes to compensation, end strength, or grade strength.

Chapter 6. Conclusion

Spending on active duty MILPERS has outpaced price growth in other areas of the economy since FY 2000, even as the size of the workforce decreased. This is a crippling problem—the workforce is the foundation for mission readiness. This is also a far-reaching problem—the increasing cost of MILPERS threatens to crowd out efforts to modernize key military capabilities. Finally, this is a highly interconnected problem—although options exist to reduce MILPERS spending, they might entail unacceptable risks throughout the DAF enterprise.

The objective of this research was to create a strategic modeling capability to estimate the cost effects from changing workforce and personnel policies. The analytic tool we created marries elements for examining how authorizations and personnel policies influence the makeup of the workforce, how the makeup of the workforce influences the average cost of an airman, and how the size of the workforce and the cost of an airman drive MILPERS spending. This report demonstrates how that capability can be used to explore options that reduce MILPERS spending.

Key Findings

The key results of our analysis of MILPERS spending and our simulations using the analytic tool are summarized in Table 6.1. Together, these support three general findings.

First, MILPERS spending and the average cost of an airman, although related, are distinct. MILPERS spending is notionally the product of end strength and the cost of an airman. Reducing end strength will limit MILPERS spending but without affecting the average cost of an airman. Reducing the average cost of an airman might also limit MILPERS spending but without reducing end strength.

Second, different options for reducing the average cost of an airman exist, for example, fielding a more-junior or less-experienced workforce, shifting work to more-economical labor categories, or limiting growth in compensation. Some of these options rely on personnel policies, such as reducing HYT or selective continuation, to drive changes in attributes of the workforce that contribute to cost. The savings from these options, although significant, tend to be smaller than the savings from changing the authorized workforce structure.

Third, all solution options entail trade-offs. For example, reducing end strength decreases the quantity of labor available, reducing growth in compensation decreases retention, shifting toward a more-junior grade mix decreases retention, and reducing retention decreases experience (i.e., YOS) and increases recruiting demands. Some of these trade-offs can be directly quantified. For instance, the increased cost of recruiting and training could erode MILPERS savings associated with fielding a more-junior workforce. Other trade-offs cannot be directly quantified. For instance, how do changes in experience level, as measured by YOS, translate to changes in ability? Harder still, how do different solution options affect morale?

Table 6.1. Summary of Key Results

Chapter	Focus	Key Results
2	Drivers of MILPERS spending	• Basic pay, BAH, and BAS account for the largest share of standard composite pay rates. Options that limit the rate of growth in these elements will reduce the average cost of an airman in the future. • Grade and YOS affect many elements of standard composite pay rates. Options that affect these workforce characteristics will change the cost of an airman in the future.
3	Historical and cross-service analysis of MILPERS spending	• Changes in standard composite pay rates explain most of the movement in the DAF's MILPERS budget from FY 2000 to FY 2021. Options that limit the rate of growth in standard composite pay rates will reduce the average cost of an airman and spending on MILPERS. • Changes in end strength have also explained some of the movement in the MILPERS budget. Options that reduce end strength will reduce MILPERS spending, potentially without affecting the average cost of an airman. • As compared with the Army and the Marines, the DAF has a higher officer-to-enlisted ratio and a more-senior grade mix. Options that reduce grade mix differences between the DAF and the other services have the potential to reduce future MILPERS costs.
4	Analytic framework for exploring options to reduce MILPERS spending	• Because of the interconnected nature of personnel policies, the workforce, and MILPERS spending, personnel policies might be used to reduce the average cost of an airman. Relatedly, personnel policies designed to achieve other objectives have the potential to contribute to MILPERS spending. • Because of the interconnected nature of the personnel system, solution options entail such trade-offs as reduced end strength, reduced experience levels, and increased recruiting and accession costs.
5	Application of analytic framework to solution options	• Certain personnel policies would allow the DAF to save $10 million to $100 million annually on MILPERS. • Shifting work to enlisted personnel and less-senior individuals might achieve annual MILPERS savings on the order of $100 million to billions annually. • Limiting annual increases in compensation could achieve annual MILPERS savings on the order of $500 million to billions annually. • Reducing end strength might achieve annual MILPERS savings on the order of billions annually. • These solution options entail different trade-offs in terms of cost, capability, and experience levels.

A limitation of our analysis of solution options is that the primary outcome, MILPERS spending, is unmoored—the magnitude of savings is not linked to the size of the workforce or policy change. The analytic tool provides a useful starting point for estimating order-of-magnitude effects and identifying promising solutions. Further analysis would be needed to evaluate implementation barriers and other costs for the most-promising subset of solutions.

Recommendations

Recommendation #1. The DAF could use a simulation capability like the one described here to link workforce and policy changes to MILPERS spending. Because of the interconnected nature of workforce and personnel policies, personnel inventory, and standard composite pay rates, analytic tools are needed to predict how solution options will affect MILPERS spending. To implement this recommendation, the DAF should:

- **Use the simulation capability to explore how personnel policies could be used to decrease MILPERS spending.** Many elements of standard composite pay rates depend on YOS. Personnel policies that reduce YOS, such as reducing HYT or selective continuation, might reduce the average cost of an airman without affecting end strength or grade mix.
- **Use the simulation capability to explore how personnel policies will affect future MILPERS spending.** Personnel policies developed to meet other objectives, such as retaining skilled servicemembers, might indirectly affect the average cost of an airman. Using a simulation capability like the one described here would allow the DAF to anticipate how these policies will affect future MILPERS spending.

Recommendation #2. Improve the fidelity and breadth of the simulation capability. The analytic tool described here establishes a basis for continuing lines of development. To implement this recommendation, the DAF should:

- **Include the ability to simulate the year-over-year dynamics produced by changes in workforce and personnel policy.** Currently, the tool projects the long-run results of policy changes. A key direction for future development is to include the capability to simulate the annual dynamics during the transition from the current to the future state of the workforce.
- **Increase the fidelity of monetary outcomes included in the analytic tool.** Certain values included in the analytic tool (e.g., training costs and the costs of individuals in different labor categories) are approximations. As reported in AFI 65-503 Table A18-A and Table A18-B, initial skills training costs do not distinguish between MILPERS costs, operation and maintenance costs, and other costs. Training costs must be reported at a more-granular level to capture the additional, non-MILPERS expense of increasing annual accessions. More-granular data about civilian costs (e.g., by general schedule level and location) are available and could immediately be incorporated into the analytic tool.
- **Include additional monetary outcomes in the analytic tool.** The analytic tool primarily describes MILPERS costs. However, solution options might drive changes in the costs of programs and activities outside the MILPERS budget, such as the Defense Health Program, DoD Education Activity, and other in-kind benefits. The analytic tool could be expanded to include these programs and activities, which amount to about 20 percent of direct personnel costs.

49

- **Include additional, nonfinancial outcomes in the analytic tool.** The analytic tool lacks a measure of mission essentiality. Therefore, such metrics as risk to mission cannot be quantified.

Recommendation #3. Refine and evaluate solution options. Several options explored in Chapter 5 showed the potential to increase MILPERS affordability. The DAF should refine and further evaluate some of these options. Specifically, the DAF should:

- **Reconsider personnel requirements for platforms, missions, and operations.** The DAF should reconsider the apportionment of responsibility between officers and enlisted personnel. In addition, the DAF should also reconsider the rank and experience necessary for certain functions. By more-fully utilizing the abilities of noncommissioned officers and more-junior personnel, the DAF could reduce MILPERS spending. However, delaying advancement by shifting to a more-junior grade mix might have negative retention effects.
- **Apply workforce and personnel policies in a targeted manner.** Policies like reducing the percentage of six-year initial service commitments for enlisted personnel, adjusting bonuses to offset deferred pay increases, and reducing HYT and selective continuation can be tailored for "recruit" versus "retain" career fields. These policies can be applied in a targeted manner to reduce overall experience (and MILPERS spending) while sustaining retention in career fields with critical shortages or high recruiting and initial training costs.

Recommendation #4. Develop solution options with input from operations, plans, programs, financial management, logistics, engineering, and force protection communities. Approaches for reducing MILPERS spending might introduce risk throughout the DAF enterprise; implementing any of these options would have secondary and tertiary consequences that must be identified and considered. The problem cannot be solved in the silo of manpower, personnel, and services. Additional perspectives from across the air staff and secretariat are needed to identify risks associated with different solution options and to develop mitigating courses of action. To implement this recommendation, the DAF should conduct recurring workforce policy games with diverse participation to develop solution options and reach consensus.

Appendix A. Decomposing Changes in the DAF MILPERS Budget over Time

Changes in MILPERS spending can be linked to changes in end strength, grade mix, and standard composite pay rates. To provide insight into the drivers of changes in MILPERS spending over time, we calculated indexes to highlight the contributions of these factors.

Framework for Decomposing MILPERS Spending over Time

To decompose the DAF's MILPERS budget in year t (B_t), we represent it as a function of the total active duty end strength in year t (E_t), the share of active duty personnel (measured in work years) in grade i in year t ($s_{i,t}$), and the standard composite pay rate associated with grade i in year t ($c_{i,t}$). Under MILPERS account rules, the total budget is

$$B_t = E_t \cdot \Sigma_i \, s_{i,t} \cdot c_{i,t} \text{ (MILPERS budget equation)},$$

where $\Sigma_i \, s_{i,t} = 1$ for all t.

Over time, end strength (E_t), grade mix ($s_{i,t}$), and standard composite pay rates ($c_{i,t}$) will change and contribute to movements in the MILPERS budget (B_t). To isolate the effect of each of these components, we developed indexes that show how much of the difference in the MILPERS budget in year t relative to FY 2021 is because of changes in end strength, grade mix, and standard composite pay rates.[73] The indexes are a means of isolating the effect of changes in each component of the MILPERS budget equation above, holding other factors constant.

The indexes we calculate are

$$I_t^{End\ Strength} = \frac{E_t \cdot \Sigma_i \, s_{i,2021} \cdot c_{i,2021}}{B_{2021}} = \frac{E_t}{E_{2021}} \text{ (end strength index)}$$

$$I_t^{Grade\ Mix} = \frac{E_{2021} \cdot \Sigma_i \, s_{i,t} \cdot c_{i,2021}}{B_{2021}} = \frac{\Sigma_i \, s_{i,t} \cdot c_{i,2021}}{\Sigma_i \, s_{i,2021} \cdot c_{i,2021}} \text{ (grade mix index)}$$

$$I_t^{Composite\ Rate} = \frac{E_{2021} \cdot \Sigma_i \, s_{i,2021} \cdot c_{i,t}}{B_{2021}} = \frac{\Sigma_i \, s_{i,2021} \cdot c_{i,t}}{\Sigma_i \, s_{i,2021} \cdot c_{i,2021}} \text{ (composite pay rate index)}$$

We also calculate a simple index of the change in MILPERS spending, which reflects the combined effect of all three factors noted above:

---

[73] The indexes can be interpreted like inflation indexes (see, for example, Chapter 5 of the DoD Green Book [OUSD, 2021c]). An increase in the index in year t for a factor is indicative of that factor contributing to an increase in MILPERS spending in year t. Each index is normalized so that it equals a value of 1.0 in the year 2021.

$$I_t^{MILPERS} = \frac{B_t}{B_{2021}} \text{ (MILPERS index)}$$

To help assess the nominal growth in standard composite pay rates captured by the composite pay rate index, we also report out indexes measuring the nominal rate of growth in civilian wages and overall prices in the U.S. economy as measured by the GDP price deflator.

Application of Framework to the DAF's MILPERS Budget

Figure A.1 shows the indexes noted above generated for the DAF's MILPERS budget. The figure suggests the following:

- Changes in standard composite pay rates for active duty personnel have historically most-closely explained movements in the DAF's MILPERS budget. Standard composite pay rates grew by approximately 4.8 percent per year between 2000 and 2012. Between 2012 and 2021, that growth slowed to 1.4 percent per year. Growth in standard composite pay rates for DAF active duty military personnel have outpaced growth in civilian wages and overall prices in the U.S. economy.
- End strength movements are the second most-important factor affecting the MILPERS budget. End strength was higher during the 2000 to 2007 time frame than in 2021, contributing to a higher MILPERS budget during that period. End strength was lower during the 2014 to 2018 time frame than in 2021, contributing to a lower MILPERS budget during that period.
- Shifts in the share of regular DAF end strength assigned to different grades have typically explained very little of the movement the DAF's MILPERS budget.

Figure A.1. Indexes of Drivers of DAF MILPERS Spending Increases

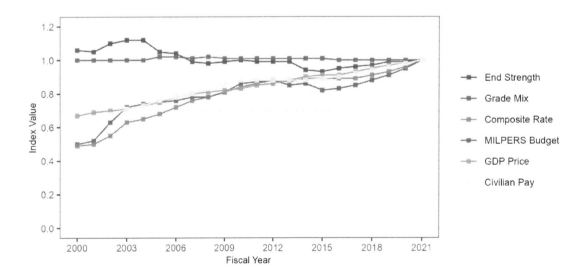

SOURCE: Data reported in DAF Military Personnel Program President Budget Submissions.

Appendix B. Description of the Manpower Realignment Tool

The analytic tool described in Chapter 4 is implemented as a Microsoft Excel workbook. The workbook contains multiple sheets described in the subsequent sections of this appendix.

- *Solution Summary*: reports outputs of the policy model run
- *Personnel Policy*: allows a user to implement personnel policies
- *Workforce Management Policy*: allows a user to implement workforce management policies
- *MPES Interface*: allows a user to apply changes to targeted segments of the workforce
- *Officer_Model (hidden)*: contains a model for simulating how personnel and workforce policies change the officer inventory
- *Enlisted_Model (hidden)*: contains a model for simulating how personnel and workforce policies change the enlisted inventory
- *MPES_Records (hidden)*: contains FY 2024–funded authorizations exports from MPES in December 2021
- *Resets (hidden)*: contains values to reset the workbook to baseline state.

Solution Summary Sheet

The solution summary sheet contains tables with simulation outputs (see Figure B.1). The portion titled "Solution Metrics" compares five metrics between the user-defined solution and a baseline solution that does not include any new personnel or workforce policies: (1) total cost, (2) cost of an airman, (3) FTE, (4) experience (YOS), and (5) total accessions. For all metrics, the total and the percentage change from the baseline is reported, and cell colors reflect the magnitude and direction of the change.

The portion titled "Officer Summary" shows detailed results for officers. The table includes four metrics by grade: (1) authorizations, as specified by workforce policies; (2) assigned, as projected by the simulation model; (3) annual composite, as calculated using AFI 65-503 and adjusting for YOS as projected by the simulation model; and (4) mean YOS, as projected by the simulation model. The table also shows the number of annual accessions needed to sustain the workforce. Once again, for all metrics, the total and the percentage change from the baseline is reported, and cell colors reflect the magnitude and direction of the change. The table titled "Enlisted Summary" shows the same detailed results for enlisted personnel.

Lastly, the portion titled "Non-RegAF" shows the number of additional civilians, contractors, and enlisted and officer reservists included in the solution.

Figure B.1. Solution Summary

Solution Metrics		Solution	Baseline	Delta	Percent
Total Cost		$33,550,188,445	$35,023,243,541	($1,473,055,097)	(-4.2%)
	Officer	$10,549,650,303	$10,732,728,958	($183,078,656)	(-1.7%)
	Enlisted	$22,061,123,204	$23,401,820,525	($1,340,697,322)	(-5.7%)
	Non-RegAF	$0	$0	$0	
	Accessions	$939,414,938	$888,694,058	$50,720,880	(5.7%)
Cost of an Airman		$98,364	$102,960	($4,596)	(-4.5%)
	Officer	$166,464	$169,353	($2,889)	(-1.7%)
	Enlisted	$82,269	$87,269	($5,000)	(-5.7%)
Full Time Equivalent		331,533	331,533	0	(0%)
	Officer	63,375	63,375	0	(0%)
	Enlisted	268,158	268,158	0	(0%)
	Non-RegAF	0	0	0	
Experience (YOS)		7.0	7.5	-0.5	(-6.6%)
	Officer	8.6	8.8	-0.2	(-2.6%)
	Enlisted	6.6	7.2	-0.6	(-7.7%)
Total Accessions		32,962	31,182	1,780	(5.7%)
	Officer	4,374	4,352	22	(0.5%)
	Enlisted	28,587	26,830	1,758	(6.6%)

Officer Summary	Authorizations (% Delta Baseline)		Assigned (% Delta Baseline)		Annual Composite (% Delta Baseline)		Mean YOS (% Delta Baseline)		Accessions (% Delta Baseline)	
GO	334	(-11.8%)	334	(-11.8%)	$ 299,555	(0%)	26.0	(0%)	4,374	(0.5%)
O-6	2,551	(-27.2%)	2,551	(-27.2%)	$ 250,833	(0%)	22.4	(-0.2%)		
O-5	7,727	(-24.2%)	7,727	(-24.2%)	$ 213,109	(0%)	16.7	(0.2%)		
O-4	15,712	(3.3%)	15,711	(3.3%)	$ 188,046	(0.6%)	12.1	(5.5%)		
O-3	24,893	(8.1%)	21,463	(14.2%)	$ 157,538	(0.9%)	6.3	(15.8%)		
O-2	12,157	(10.1%)	8,918	(2.9%)	$ 128,314	(0%)	2.3	(-0.7%)		
O-1			6,670	(0.7%)	$ 102,373	(0%)	0.5	(0%)		
Total	63,375	(0%)	63,375	(0%)	166,463.9		8.57			

Enlisted Summary	Authorizations (% Delta Baseline)		Assigned (% Delta Baseline)		Annual Composite (% Delta Baseline)		Mean YOS (% Delta Baseline)		Accessions (% Delta Baseline)	
E-9	3,208	(20.7%)	3,206	(20.6%)	$ 153,852	(-0.1%)	21.1	(-0.3%)	28,587	(6.6%)
E-8	8,047	(44.3%)	8,047	(44.3%)	$ 134,606	(-0.1%)	19.1	(-0.6%)		
E-7	17,278	(-38.4%)	17,278	(-38.4%)	$ 121,787	(-0.3%)	17.0	(-2.3%)		
E-6	31,282	(-34.8%)	31,282	(-34.8%)	$ 106,698	(0.4%)	13.1	(3.7%)		
E-5	52,581	(-31.6%)	52,581	(-28.5%)	$ 90,837	(2%)	7.8	(21.7%)		
E-4	72,192	(32.3%)	72,192	(32.3%)	$ 74,686	(2.4%)	4.0	(59.2%)		
E-3	83,570	(59.3%)	70,748	(61.6%)	$ 59,952	(1.2%)	2.0	(77.2%)		
E-2			6,249	(6.6%)	$ 52,395	(0%)	1.0	(0%)		
E-1			6,575	(6.6%)	$ 46,515	(0%)	0.0	(0%)		
Total	268,158	(0%)	268,158	(0%)	$ 82,269		6.6			

Non-RegAF	Total
Civilian	0
Contractor	0
Enlisted ARC (Drill)	0
Officer ARC (Drill)	0

SOURCE: Screenshot from the analytic tool.

Personnel Policy Sheet

The personnel policy sheet allows the user to implement various personnel policies (see Figure B.2). The user might adjust the start of promotion windows or HYT for officers and enlisted personnel, as well as the width of the promotion window. The user might also modify the percentage of enlisted personnel entering the workforce with advanced rank (E-2 or E-3). In addition, the user might modify military pay relative to civilian wages. Finally, the user might change the retention elasticity associated with change in pay.

After specifying new personnel policies, the user must run the simulation models to find promotion rates and to generate inventory projections that reflect the changes.

Figure B.2. Personnel Policy Tab

Rank	Start of Promotion Window (Year)	High Year of Tenure	Length of Promotion Window
GO	23	40	1
O-6	20	30	1
O-5	14	28	1
O-4	9	24	1
O-3	4	20	1
O-2	2	5	1
O-1	0	3	1

Source: 10 U.S. Codes 631-634

Rank	Start of Promotion Window (Year)	High Year of Tenure	Length of Promotion Window
E-9	19	30	10
E-8	17	26	10
E-7	15	24	10
E-6	11	22	10
E-5	6	20	5
E-4	3	10	5
E-3	2	8	1
E-2	1	8	1
E-1	0	8	1

Source: 30 Oct 2019 AF Memo

% Airmen Entering at Rank of E-2/E-3	77.0%
% Pay Change Relative to Civilian	0.0%
% Change in Retention per 1% Change in Pay	1.2%

SOURCES: Screenshot from the analytic tool; 10 U.S.C. 631; 10 U.S.C. 632; 10 U.S.C. 633; 10 U.S.C. 634.

Force Management Policy Sheet

The force management policy sheet allows the user to apply changes to workforce structure and mix (see Figure B.3). The user could directly change grade mix in the tables titled "Authorized Officer Pyramid" and "Authorized Enlisted Pyramid." The user could change end strength by AFSC in the tables labeled "Authorized Officer Career Field Strength" and "Authorized Enlisted Career Field Strength."

After specifying new workforce policies, the user must run the simulation models to find promotion rates and to generate inventory projections that reflect the changes.

Figure B.3. Force Management Policy

Authorized Officer Pyramid

Rank	Baseline	Solution
GO	0.6%	0.6%
O-6	5.5%	5.5%
O-5	16.1%	16.1%
O-4	24.0%	24.0%
O-3	36.4%	36.4%
O-2	17.4%	17.4%
O-1		
Total		100.0%

Authorized Enlisted Pyramid

Rank	Baseline	Solution
E-9	1.0%	1.0%
E-8	2.1%	2.1%
E-7	10.5%	10.5%
E-6	17.9%	17.9%
E-5	28.7%	28.7%
E-4	20.4%	20.4%
E-3	19.6%	19.6%
E-2		
E-1		
Total		100.0%

Authorized Officer Career Field Strength

AFSC Title	AFSC	Baseline	Solution
OPERATIONS COM	10Cx	313	313
BOMBER PILOT	11Bx	665	665
EXPER TEST PILOT	11Ex	181	181
FIGHTER PILOT	11Fx	3,071	3,071
GENERALIST PILOT	11Gx	359	359
RESCUE PILOT	11Hx	760	760
TRAINER PILOT	11Kx	1,708	1,708
MOBILITY PILOT	11Mx	3,776	3,776
RECON SURVEIL E\	11Rx	656	656
SPECIAL OPS PILOT	11Sx	1,280	1,280
RMTLY PILOT ACFT	11Ux	14	14

Authorized Enlisted Career Field Strength

AFSC Title	AFSC	Baseline	Solution
IN-FLIGHT REFUEL	1A0x	825	825
FLIGHT ENGINEER	1A1x	521	521
AIRCRAFT LOADM/	1A2x	2,022	2,022
AIRBORNE MISSIO	1A3x	1,872	1,872
FLIGHT ATTENDAN	1A6x	286	286
ABN CRYPT LANG /	1A8x	1,572	1,572
SPC MISSION AVIA	1A9x	957	957
CYBER WARFARE (	1B0x	12	12
CYBER WARFARE (	1B4x	1,832	1,832
AVIATION RESOUR	1C0x	1,956	1,956
AIR TRAFFIC CONT	1C1x	2,584	2,584

SOURCE: Screenshot from the analytic tool.

The force management policy sheet gives the user limited to ability to reallocate work to different labor categories (see Figure B.4). Specifically, the user can increase the total number of civilians, contracts, and reservists and can convert officer positions to enlisted positions.

Figure B.4. Changes in Labor Categories

Non-RegAF

	Total		Annual Composite	Conversion	FTE
Civilian	0	$	122,000	1	1
Contractor	0	$	200,000	1	1
Enlisted ARC (Drill)	0	$	21,000	1	0.1
Officer ARC (Drill)	0	$	47,000	1	0.1
	$	-			

Conversions

	Total
Officer Losses	0
Enlisted Conversions	0

SOURCE: Screenshot from the analytic tool.

MPES Interface

The MPES interface allows the user to apply targeted changes in end strength levels to workforce segments (see Figure B.5). For example, to reduce the number of authorizations in the tail of the workforce, the user selects the corresponding level for that attribute and enters a reduction percentage. The results of the change are automatically passed into the force management policy sheet.

After applying a change, the user must run the simulation models to find promotion rates and to generate inventory projections that reflect the change.

Figure B.5. MPES Interface

Variable	OFFICER_ENLISTED	GRADE	AFSC	AFSC_SERIES	ACQUISITION_OR_STEM	TOOTH_OR_TAIL	WING_LEVEL
Percent Reduction	0.0%	0.0%	0.0%	0.0%	0.0%	0.0%	0.0%
View Levels						TOOTH	

SOURCE: Screenshot from the analytic tool.

Abbreviations

AFI	Air Force Instruction
AFSC	Air Force Specialty Code
ARC	Air Reserve Component
BAH	basic allowance for housing
BAS	basic allowance for subsistence
CBO	Congressional Budget Office
COA	course of action
CSAF	chief of staff of the Air Force
DAF	Department of the Air Force
DoD	U.S. Department of Defense
ECI	Employment Cost Index
FCoM	Full Cost of Manpower
FTE	full-time equivalent
FY	fiscal year
GDP	gross domestic product
GO	general officer
HYT	high year of tenure
IPZ	in the promotion zone
MERHC	Medicare-Eligible Retiree Health Care Accrual
MILPERS	military personnel
MPES	Manpower Programming and Execution System
Non-RegAF	nonregular Air Force personnel
PCS	permanent change of station
RPA	retirement pay accrual
SAF/FMB	Deputy Assistant Secretary for Budget
STEM	science, technology, engineering, and mathematics
TIG	time in grade
YOS	year(s) of service

References

AFI—*See* Department of the Air Force Instruction.

Brown, Charles Q., *CSAF Action Orders to Accelerate Change Across Air Force*, Arlington, Va.: Chief of Staff, U.S. Air Force, December 2020.

CBO—*See* Congressional Budget Office.

Congressional Budget Office, *Replacing Military Personnel in Support Positions with Civilian Employees*, Washington, D.C., December 2015. As of February 7, 2022:
https://www.cbo.gov/publication/51012

———, *Approaches to Changing Military Compensation*, Washington, D.C., January 2020. As of February 7, 2022:
https://www.cbo.gov/publication/55648

———, *Alternative Approaches to Adjusting Military Cash Pay*, September 2021. As of February 7, 2022:
https://www.cbo.gov/publication/57192

Correll, Diana S., "Air Force Extends First-Term, Unaccompanied Tours at Some Overseas Duty Stations to 36 Months," *Air Force Times*, February 12, 2021.

DAF—*See* Department of the Air Force.

Daniels, Seamus P., "Assessing Trends in Military Personnel Costs," Center for Strategic and International Studies, September 2021. As of February 7, 2022:
https://www.csis.org/analysis/assessing-trends-military-personnel-costs

"Defense Primer: Military Enlisted Personnel," Washington, D.C.: Congressional Research Service, updated December 1, 2021.

"Defense Primer: Military Pay Raise," Washington, D.C.: Congressional Research Service, updated December 27, 2021.

Defense Travel Management Office, "Allowances," webpage, undated. As of February 7, 2022:
https://www.defensetravel.dod.mil/site/allowances.cfm

Department of the Air Force, *Fiscal Year (FY) 2023 Military Personnel Appropriation*, Washington, D.C., April 2022.

Department of the Air Force, *FY 2002 Amended Budget Submission*, June 2001.

Department of the Air Force, "Memorandum for MAJCOM/CCs, Field Command/CCs, Input Sources, and Air Force Corporate Structure Members: Addendum 23-014 (Functional Optimization for Affordability)," Washington, D.C., September 2020.

Department of the Air Force Instruction 65-503, *Financial Management: US Air Force Cost and Planning Factors*, Washington, D.C., July 13, 2018.

Department of the Air Force Instruction 36-2606, *Reenlistment and Extension of Enlistment in the United States Air Force*, Washington, D.C., September 20, 2019a.

Department of the Air Force Instruction 36-2502, *Enlisted Airman Promotion and Demotion Programs*, Washington, D.C., September 27, 2019b.

Department of the Air Force Instruction 36-2032, *Military Recruiting and Accessions*, Washington, D.C., September 27, 2019c.

———, "Memorandum for MAJCOM/CCs, Field Command/CCs, Input Sources, and Air Force Corporate Structure Members: Addendum 23-014 (Functional Optimization for Affordability)," September 2020.

Department of the Air Force Instruction 36-2501, *Officer Promotion and Selective Continuation*, Washington, D.C., April 30, 2021.

Deputy Chief of Staff for Manpower, Personnel and Services, "Accelerate *Workforce* Change or Lose: An A1 Addendum to *CSAF's Accelerate Change or Lose*," June 2021.

DoD—*See* U.S. Department of Defense.

Harrison, Todd, "Rethinking the Role of Remotely Crewed Systems in the Future Force," Center for Strategic and International Studies, March 3, 2021.

Hosek, James, Beth J. Asch, and Michael G. Mattock, *Toward Efficient Military Retirement Accrual Charges*, Santa Monica, Calif.: RAND Corporation, RR-1373-A, 2017. As of February 3, 2022:
https://www.rand.org/pubs/research_reports/RR1373.html

Menthe, Lance, Dahlia Anne Goldfeld, Abbie Tingstad, Sherrill Lingel, Edward Geist, Donald Brunk, Amanda Wicker, Sarah Soliman, Balys Gintautus, Anne Stickells, and Amado Cordova, *Technology Innovation and the Future of Air Force Intelligence Analysis: Volume 1, Findings and Recommendations*, Santa Monica, Calif.: RAND Corporation, RR-A341-1, 2022. As of February 7, 2022:
https://www.rand.org/pubs/research_reports/RRA341-1.html

Office of the Secretary of Defense, Office of Cost Assessment and Program Evaluation, *FCOM Military Rates 2017: White Paper—References, Calculations, and Assumptions*, Washington, D.C.: U.S. Department of Defense, April 2017.

Office of the Secretary of Defense (Comptroller), "FY 2021 Department of Defense Military Personnel Composite Standard Pay and Reimbursement Rates," May 11, 2020.

Office of the Under Secretary of Defense (Comptroller), *Fiscal Year (FY) 2021 Department of Defense (DoD) Fixed Wing and Helicopter Reimbursement Rates*, Washington, D.C., October 2020.

—————, *Department of Defense Financial Management Regulation (DoD FMR)*, Vol. 11A, "Reimbursable Operations Policy," Chapter 6, Appendix I, Washington, D.C., DoD 7000.14-R, updated May 2021a.

—————, *Department of Defense Budget Fiscal Year 2021: Operation and Maintenance Programs (O-1) Revolving and Management Funds (RF-1)*, Washington, D.C., May 2021b. As of February 7, 2022:
https://comptroller.defense.gov/Budget-Materials/

—————, *National Defense Budget Estimates for Fiscal Year 2022*, (Green Book), Washington, D.C., August 2021c.

OUSD—*See* Office of the Under Secretary of Defense (Comptroller).

Robbert, Albert A., Lisa M. Harrington, Louis T. Mariano, Susan A. Resetar, David Schulker, John S. Crown, Paul Emslie, Sean Mann, and Gary Massey, *Air Force Manpower Determinants: Options for More-Responsive Processes*, Santa Monica, Calif.: RAND Corporation, RR-4420-AF, 2020. As of January 20, 2022:
https://www.rand.org/pubs/research_reports/RR4420.html

Secretary of the Air Force Public Affairs, U.S. Air Force, "Air Force Extends High Year of Tenure for E-4s Through E-6s," Washington, D.C.: U.S. Air Force, October 22, 2018. As of February 7, 2022:
https://www.af.mil/News/Article-Display/Article/1668296/air-force-extends-high-year-of-tenure-for-e-4s-through-e-6s/

—————, "Air Force Announces NCO Career Status Program for Airmen with 12 Years of Service," Washington, D.C.: U.S. Air Force, October 30, 2019. As of September 19, 2022:
https://www.af.mil/News/Article-Display/Article/2003354/air-force-announces-nco-career-status-program-for-airmen-with-12-years-of-servi/

U.S. Air Force, "Air Force Demographics," webpage, Air Force's Personnel Center, undated. As of February 8, 2022:
https://www.afpc.af.mil/About/Air-Force-Demographics/

U.S. Air Force, Financial Management and Comptroller, "Air Force President's Budget FY22," May 2021. As of February 7, 2022:
https://www.saffm.hq.af.mil/FM-Resources/Budget/Air-Force-Presidents-Budget-FY22

U.S. Code, Title 10, Section 631, Effect of Failure of Selection for Promotion: First Lieutenants and Lieutenants (Junior Grade), 2010.

U.S. Code, Title 10, Section 632, Effect of Failure of Selection for Promotion: Captains and Majors of the Army, Air Force, and Marine Corps and Lieutenants and Lieutenant Commanders of the Navy, 2011.

U.S. Code, Title 10, Section 633, Retirement for Years of Service: Regular Lieutenant Colonels and Commanders, 2010.

U.S. Code, Title 10, Section 634, Retirement for Years of Service: Regular Colonels and Navy Captains, 2015.

U.S. Department of Defense, "Military Compensation—Special and Incentive Pay," website, undated. As of February 7, 2022:
https://militarypay.defense.gov/Pay/Special-and-Incentive-Pays/

———, *Report of the Ninth Quadrennial Review of Military Compensation*, Vol. 1, Washington, D.C., March 2002.

———, *Report of the Thirteenth Quadrennial Review of Military Compensation*, Washington, D.C., December 2020.

U.S. Government Accountability Office, "Military Compensation: DoD Needs More Complete and Consistent Data to Assess the Costs of Policies on Relocating Personnel," Washington, D.C., GAO-15-173, September 2015. As of February 7, 2022:
https://www.gao.gov/products/gao-15-713

Walsh, Matthew, David Schulker, Nelson Lim, Albert A. Robbert, Raymond E. Conley, John S. Crown, and Christopher E. Maerzluft, *Department of the Air Force Officer Talent Management Reforms: Implications for Career Field Health and Demographic Diversity*, Santa Monica, Calif.: RAND Corporation, RR-A556-1, 2021. As of February 7, 2022:
https://www.rand.org/pubs/research_reports/RRA556-1.html

Warner, John T., "The Effect of the Civilian Economy on Recruiting and Retention," in U.S. Department of Defense, *Report of the Eleventh Quadrennial Review of Military Compensation*, supporting research papers, Part 1, Chapter 2, June 2012. As of February 7, 2022:
https://go.usa.gov/xVBxq